新时代 新工匠
职业教育改革创新系列教材

网络信息安全一体化教程

胡定松 黄四清 主 编
周允强 朱国艺 副主编

电子工业出版社
Publishing House of Electronics Industry
北京·BEIJING

内 容 简 介

本书贯彻基于工作过程的课程理念，以"项目式教学"为主要思想，建立以项目为核心、以工作过程为导向、以真实的工作任务为驱动机制的教学过程，采用教、学、做一体化的方式撰写，合理地组织任务，并将每个任务分解为需求分析、方案设计、知识准备、任务实现四个模块，体现教、学、做一体化过程。

全书分为十个项目，内容包括设备安全、VLAN 技术与生成树技术、路由协议、路由重分布与策略路由、访问控制列表、DHCP 与 VRRP、组播、广域网技术、网络安全与 VPN 实现、IPv6 技术与实施。

本书可以作为职业院校、技术院校计算机及相关专业计算机网络课程的实验教材，也可以作为网络培训或相关工程技术人员自学的参考书。

未经许可，不得以任何方式复制或抄袭本书之部分或全部内容。
版权所有，侵权必究。

图书在版编目（CIP）数据

网络信息安全一体化教程 / 胡定松，黄四清主编. —北京：电子工业出版社，2017.7
ISBN 978-7-121-32241-9

Ⅰ. ①网… Ⅱ. ①胡… ②黄… Ⅲ. ①计算机网络—信息安全—教材 Ⅳ. ①TP393.08

中国版本图书馆 CIP 数据核字（2017）第 169097 号

策划编辑：关雅莉
责任编辑：柴 灿
印　　刷：北京七彩京通数码快印有限公司
装　　订：北京七彩京通数码快印有限公司
出版发行：电子工业出版社
　　　　　北京市海淀区万寿路 173 信箱　邮编　100036
开　　本：787×1 092　1/16　印张：13　字数：382 千字
版　　次：2017 年 7 月第 1 版
印　　次：2023 年 8 月第 5 次印刷
定　　价：28.00 元

凡所购买电子工业出版社图书有缺损问题，请向购买书店调换。若书店售缺，请与本社发行部联系，联系及邮购电话：（010）88254888，88258888。
质量投诉请发邮件至 zlts@phei.com.cn，盗版侵权举报请发邮件至 dbqq@phei.com.cn。
本书咨询联系方式：（010）88254617，luomn@phei.com.cn。

前　言

本书是首批国家级中等职业教育改革与发展示范学校重点建设专业"计算机网络技术"的成果之一，在调研本区域内相关企业岗位的职业能力基础上，深入分析和提取网络技术专业的典型工作任务和岗位技能目标，同时依据毕业生跟踪调查的数据分析，逐步探索、形成新的人才培养方案和专业课程体系，特别是在一体化课程建设方面，取得了预期的目标。

在编写思想上，本书贯彻"基于工作过程"的课程理念，以"项目式教学"为主要思想，建立以项目为核心、以工作过程为导向、以真实的工作任务为驱动机制的教学过程。采用教、学、做一体化的方式进行撰写，合理地组织任务，并将每个任务分解为需求分析、方案设计、知识准备、任务实现四个模块，体现教、学、做一体化的过程。本书侧重于实用性强的网络安全技术，抛开了深奥的理论，以过程化的图片和文字进行表述，直接面向实际工作中的应用环境，使读者更加直观地了解攻击或者防御手法，更加有利于促进计算机专业的学生迅速向网络安全管理人员的角色迈进。本书在内容编排上简单、直观、循序渐进，方便学生积累经验，迅速拉近理论与实践的距离。

本书精选大量的实用案例，循序渐进地介绍了计算机信息安全的基本原理及其应用技术；注重结合经典实例来讲解一些关键技术和应用难点，侧重实用性和启发性。全书分为十个项目，内容包括设备安全、VLAN 技术与生成树技术、路由协议、路由重分布与策略路由、访问控制列表、DHCP 与 VRRP、组播、广域网技术、网络安全与 VPN 实现、IPv6 技术与实施。

本书教学学时建议为 72 学时，在教学过程中可参考以下课时分配表。

项目	课程内容	课程分配		
		讲授	实训	合计
项目一	设备安全	2	2	4
项目二	VLAN 技术与生成树技术	2	4	6
项目三	路由协议	2	4	6
项目四	路由重分布与策略路由	4	6	10
项目五	访问控制列表	2	6	8
项目六	DHCP 与 VRRP	2	6	8
项目七	组播	4	8	12
项目八	广域网技术	2	4	6
项目九	网络安全与 VPN 实现	4	4	8
项目十	IPv6 技术实施	2	2	4
合计		26	46	72

本书由中山市中等专业学校胡定松、黄四清担任主编，周允强、朱国艺担任副主编。教材中的项目一～项目四由胡定松编写；项目五、项目六由黄四清编写；项目七、项目八由朱国艺编写；项目九、项目十由周允强编写。全书由胡定松统稿。

感谢神州数码网络（北京）有限公司在编者编写本书过程中给予的大力支持与指导。

由于编者水平所限，加之时间仓促，书中难免存在疏漏和不足之处，敬请广大读者批评指正。

编　者

2017 年 5 月

目　录

项目一　设备安全 ·· 1
  任务一　交换机远程管理 ·· 1
  任务二　路由器远程管理 ·· 6
  任务三　交换机端口监听 ·· 8
  任务四　交换机链路聚合 ··· 10
  任务五　交换机系统升级与备份 ·· 14
  任务六　路由器系统升级与备份 ·· 19
  认证考核 ··· 23

项目二　VLAN 技术与生成树技术 ··· 24
  任务一　实现跨交换机相同 VLAN 内通信 ·· 24
  任务二　实现不同 VLAN 间通信 ·· 28
  任务三　单实例生成树 ··· 33
  任务四　多实例生成树 ··· 37
  任务五　改变生成树状态 ··· 42
  认证考核 ··· 46

项目三　路由协议 ·· 48
  任务一　实现静态路由 ··· 48
  任务二　实现 RIP 基本配置 ·· 51
  任务三　实现 RIPv1 与 RIPv2 的兼容 ·· 55
  任务四　实现 OSPF 单区域配置 ··· 64
  任务五　实现 OSPF 多区域配置 ··· 68
  任务六　实现 OSPF 虚链路配置 ··· 71
  任务七　实现 OSPF 路由汇总 ·· 74
  任务八　实现 OSPF 认证配置 ·· 77
  认证考核 ··· 80

项目四　路由重分布与策略路由 ·· 81
  任务一　静态路由和 RIP 路由的重分布 ·· 81
  任务二　RIP 和 OSPF 的重分布 ·· 87
  任务三　基于源地址的策略路由 ·· 90
  任务四　基于应用的策略路由 ··· 95
  认证考核 ·· 100

项目五　访问控制列表 ··· 101
  任务一　标准 ACL ·· 101

任务二　扩展 ACL ………………………………………………………………… 108
　　任务三　使用 ACL 过滤特定病毒报文 …………………………………………… 113
　　认证考核 …………………………………………………………………………… 114

项目六　DHCP 与 VRRP ……………………………………………………………… 117

　　任务一　DHCP 服务器的配置 ……………………………………………………… 117
　　任务二　DHCP 中继功能的配置 …………………………………………………… 120
　　任务三　实现 VRRP 配置 …………………………………………………………… 122
　　认证考核 …………………………………………………………………………… 125

项目七　组播 …………………………………………………………………………… 127

　　任务一　使用 DVMRP 实现交换机组播的三层对接 …………………………… 127
　　任务二　使用 PIM 实现交换机组播三层对接 …………………………………… 130
　　任务三　交换机组播二层对接 …………………………………………………… 134
　　认证考核 …………………………………………………………………………… 137

项目八　广域网技术 …………………………………………………………………… 138

　　任务一　路由器串口 PPP PAP 认证 ……………………………………………… 138
　　任务二　路由器串口 PPP CHAP 认证 …………………………………………… 141
　　任务三　实现网络地址转换 ……………………………………………………… 144
　　认证考核 …………………………………………………………………………… 147

项目九　网络安全与 VPN 实现 ……………………………………………………… 149

　　任务一　路由器使用 PPTP 实现 VPDN …………………………………………… 149
　　任务二　使用 L2TP 连接企业总部与分支机构 ………………………………… 151
　　任务三　防火墙初级管理 ………………………………………………………… 155
　　任务四　防火墙典型环境安全策略实施 ………………………………………… 158
　　任务五　防火墙 SSL VPN 实现 …………………………………………………… 162
　　任务六　建立路由器 IPSec VPN 隧道 …………………………………………… 169
　　任务七　防火墙 IPSec VPN 隧道的建立 ………………………………………… 173
　　认证考核 …………………………………………………………………………… 180

项目十　IPv6 技术与实施 …………………………………………………………… 181

　　任务一　IPv6 邻居发现 …………………………………………………………… 181
　　任务二　IPv6 ISATAP 隧道搭建 ………………………………………………… 184
　　任务三　实现 6 to 4 隧道 ………………………………………………………… 186
　　任务四　IPv6 RIPng 配置 ………………………………………………………… 189
　　任务五　IPv6 OSPFv3 配置 ……………………………………………………… 197
　　认证考核 …………………………………………………………………………… 201

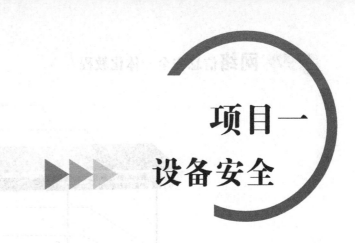

项目一 设备安全

♂ 教学背景

企事业单位在组网初期部署了设备的接入功能,随着业务的开展,员工和网络设备越来越多,渐渐地给网管人员带来了很多烦恼。由于公司网络没有经过细致规划,公司员工在各个网络接口均能上网,用户接入网络的身份无法确定,经常发现陌生的主机接入,这给公司的信息安全带来了隐患,而网络中随时可能出现的各种攻击行为也严重威胁到了网络安全。为此,需要网络管理员对设备和网络安全做出规划。

任务一 交换机远程管理

♂ 需求分析

某学校有 20 台交换机支撑着校园网的运营,这 20 台交换机分别放置在学校的不同位置。网络管理员需要对这 20 台交换机做管理。管理员可以通过带外管理的方式,即通过 Console 口来管理,但管理员需要带着自己的笔记本式计算机,并且带着 Console 线到学校的不同位置调试交换机,十分麻烦。

♂ 方案设计

校园网既然是互连互通的,在网络的任何一个信息点都应该能访问其他的信息点,为什么不通过网络的方式来调试交换机呢?通过 Telnet 方式,管理员即可在办公室中调试全校所有的交换机。

所需设备如图 1-1-1 所示。
(1)DCS 二层交换机 1 台。
(2)PC 1 台。
(3)Console 线 1 条。
(4)直通网线 1 条。

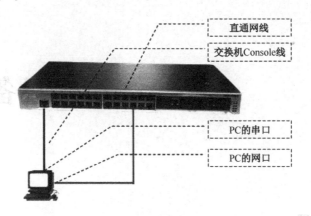

图 1-1-1　交换机远程管理

任务要求如下。
（1）按照拓扑图连接网络。
（2）PC 和交换机的 24 口用网线相连。
（3）交换机的管理 IP 地址为 192.168.1.100/24。
（4）PC 网卡的 IP 地址为 192.168.1.101/24。

♂ 知识准备

（1）默认情况下，交换机所有端口都属于 VLAN1，因此通常把 VLAN1 作为交换机的管理 Vlan，因此 VLAN1 接口的 IP 地址就是交换机的管理地址。
（2）密码只能是 1～8 个字符。
（3）删除 Telnet 用户时可以在 config 模式下使用 no telnet-user 命令。
（4）使用 Telnet 和 Web 方式调试有以下两个相同的前提条件。
① 交换机开启该功能并设置用户。
② 交换机和主机之间要能连通。
（5）有时候交换机的地址配置正确，主机配置也正确，但就是连不通。排除硬件问题之后可能的原因是主机的 Windows 操作系统开启了防火墙，关闭防火墙即可。

Telnet 方式和 Web 方式都是交换机的带内管理方式。

提供带内管理方式可以使连接在交换机中的某些设备具备管理交换机的功能。当交换机的配置出现变更，导致带内管理失效时，必须使用带外管理对交换机进行配置管理。

Web 方式也称 HTTP 方式，和 Telnet 方式一样，管理员在办公室中即可调试全校所有的交换机。

Web 方式比较简单，如果用户不习惯 CLI 的调试，则可以采用 Web 方式调试。

主流的调试界面是 CLI，大家要着重学习 CLI。

本任务使用 DCS-3926S 系列交换机作为演示设备，其软件版本为 DCS-3926S_6.1.12.0，实际使用中由于软件版本不同，功能和配置方法有可能存在差异，请关注相应版本的使用说明。

♂ 任务实现

步骤 1：给交换机的默认 VLAN 设置 IP 地址，即管理 IP 地址。

```
DCS-3926S#config
DCS-3926S(Config)#interface vlan 1                    //进入 VLAN 1 接口
02:20:17: %LINK-5-CHANGED: Interface Vlan1, changed state to UP
DCS-3926S(Config-If-Vlan1)#ip address 192.168.1.100 255.255.255.0 //配置地址
DCS-3926S(Config-If-Vlan1)#no shutdown                //激活 VLAN 接口
DCS-3926S(Config-If-Vlan1)#exit
DCS-23926S(Config)#exit
DCS-3926S#
```

验证配置：

```
DCS-3926S#show run
Current configuration:
!
hostname DCS-3926S
!
Vlan 1
vlan 1
!
Interface Ethernet0/0/1
……
Interface Ethernet0/0/24
!
interface Vlan1
interface vlan 1
ip address 192.168.1.100 255.255.255.0        //已经配置好交换机的 IP 地址
!
DCS-3926S#
```

步骤 2：为交换机设置授权 Telnet 用户。

```
DCS-3926S#config
DCS-3926S(Config)#telnet-user dcnu password 0 digital
DCS-3926S(Config)#exit
DCS-3926S#
```

步骤 3：验证配置。

```
DCS-3926S#show run
Current configuration:
!
hostname DCS-3926S
!
telnet-user dcnu password 0 digital
!
Vlan 1
vlan 1
!
Interface Ethernet0/0/1
……
```

```
Interface Ethernet0/0/24
!
interface Vlan1
interface vlan 1
ip address 192.168.1.100 255.255.255.0
!
DCS-3926S#
```

步骤 4：配置主机的 IP 地址，主机的 IP 地址要与交换机的 IP 地址在一个网段，如图 1-1-2 所示。

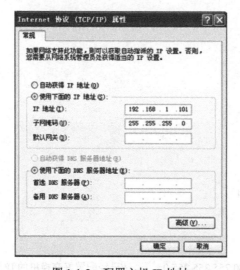

图 1-1-2　配置主机 IP 地址

步骤 5：验证配置。在主机的命令行窗口中使用 ipconfig 命令查看 IP 地址配置情况，如图 1-1-3 所示。

图 1-1-3　查看主机 IP 地址

步骤 6：验证主机与交换机是否连通。

```
DCS-3926S#ping 192.168.1.101
Type ^c to abort.
Sending 5 56-byte ICMP Echos to 192.168.1.101, timeout is 2 seconds.
```

```
!!!!!
Success rate is 100 percent (5/5), round-trip min/avg/max = 1/1/1ms
DCS-3926S#
//出现 5 个"!"表示已经连通
```

步骤 7：使用 Telnet 方式登录。登录 PC，选择"开始"→"运行"选项，弹出"运行"对话框，如图 1-1-4 所示，运行 Windows 自带的 Telnet 客户端程序，并且指定 Telnet 的目的地址，需要输入正确的登录名和口令，登录名是 dcnu，口令是 digital。

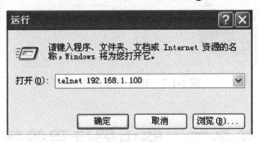

图 1-1-4　运行 Telnet 命令

步骤 8：启动交换机的 Web 服务。

```
DCS-3926S#config
DCS-3926S(Config)#ip http server          //开启 HTTP 功能
web server is on                          //表明已经成功启动
DCS-3926S(Config)#
```

步骤 9：设置交换机授权 HTTP 用户。

```
DCS-3926S(Config)#web-user admin password 0 digital    //设置密码
DCS-3926S(Config)#
```

步骤 10：使用 HTTP 方式登录。登录 PC，选择"开始"→"运行"选项，弹出"进行"对话框，如图 1-1-5 所示，指定目标。需要输入正确的登录名和口令，登录名是 admin，口令是 digital，如图 1-1-6 所示。

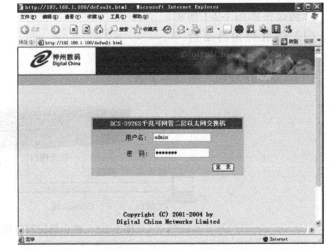

图 1-1-5　运行 HTTP 命令　　　　　　　图 1-1-6　输入用户名和密码

步骤 11：图 1-1-7 所示为交换机的 Web 调试的主界面。

图 1-1-7　Web 调试的主界面

任务二　路由器远程管理

需求分析

小张是公司的网络管理员，有时需要到不同的地方对设备进行调试，但每次都要通过计算机连接到网络设备的 Console 口进行调试，这样管理非常麻烦，小张想使用远程管理的方法来对公司的设备进行管理，这样既方便又高效。

方案设计

小张通过 Console 口管理设备，需要带着笔记本式计算机或专门设置一台台式计算机，并带着 Console 线来调试网络设备，十分麻烦。通过 Telnet 方式，小张可以坐在办公室中调试公司的所有网络设备。

所需设备如图 1-2-1 所示。

（1）DCR 路由器 2 台。

（2）PC 1 台。

（3）Console 线缆、网线各 1 条。

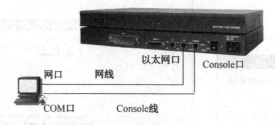

图 1-2-1　路由器远程管理

任务要求：DCR-1702 的 Console 口与 PC 的 COM 口使用 Console 线连接；F0/0 与 PC 的网卡使用交叉双绞线连接，并分别配置 192.168.2.1 和 192.168.2.2 的 C 类 IP 地址。

知识准备

（1）超级终端中的配置是对路由器的操作，此时的 PC 只是输入输出设备。
（2）在使用 Telnet 和 Web 方式管理时，先测试连通性。

任务实现

步骤 1：设置路由器以太网接口 IP 地址并验证连通性。

```
Router>enable                                              //进入特权模式
Router #config                                             //进入全局配置模式
Router-A_config#interface f0/0                             //进入接口模式
Router-A_config_f0/0#ip address 192.168.2.1 255.255.255.0  //设置 IP 地址
Router-A_config_f0/0#no shutdown
Router-A_config_f0/0#^Z
Router-A#show interface f0/0                               //验证
FastEthernet0/0 is up, line protocol is up                 //接口和协议都必须 up
address is 00e0.0f18.1a70
    Interface address is 192.168.2.1/24
    MTU 1500 bytes, BW 100000 kbit, DLY 10 usec
    Encapsulation ARPA, loopback not set
Keepalive not set
    ARP type: ARPA, ARP timeout 04:00:00
    60 second input rate 0 bits/sec, 0 packets/sec!
    60 second output rate 6 bits/sec, 0 packets/sec!
    Full-duplex, 100Mb/s, 100BaseTX, 1 Interrupt
        0 packets input, 0 bytes, 200 rx_freebuf
        Received 0 unicasts, 0 lowmark, 0 ri, 0 throttles
        0 input errors, 0 CRC, 0 framing, 0 overrun, 0 long
        1 packets output, 46 bytes, 50 tx_freebd, 0 underruns
        0 output errors, 0 collisions, 0 interface resets
        0 babbles, 0 late collisions, 0 deferred, 0 err600
        0 lost carrier, 0 no carrier 0 grace stop 0 bus error
    0 output buffer failures, 0 output buffers swapped out
```

步骤 2：设置 PC 的 IP 地址并测试连通性，如图 1-2-2 和图 1-2-3 所示。

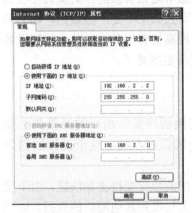

图 1-2-2　设置主机 IP 地址

图 1-2-3　测试连通性

步骤 3：设置本地数据库中的用户名，本例使用用户名 dcnu 和密码 dcnu。

Router-A_config#username dcnu password dcnu //设置本地用户名和密码
Router-A_config#

步骤 4：创建一个新的登录验证方法，名为 login_fortelnet，此方法将使用本地数据库验证。

Router-A_config#aaa authentication login login_fortelnet local
//创建 login_fortelnet 验证，采用 local
Router-A_config#

步骤 5：进入 Telnet 进程管理配置模式，配置登录用户使用 login_fortelnet 的验证方法进行验证。

Router-A_config#line vty 0 4
Router-A_config_line#login authentication login_fortelnet　　//在接口下应用
Router-A_config_line#

步骤 6：经过配置，Telnet 登录路由器时的过程如下所示。

C:\>telnet 192.168.2.1
Connecting to remote host...
Press 'q' or 'Q' to quit connection.
User Access Verification
Username:dcnu
Password:
2004-1-1 04:21:34 User dcnu logged in from 192.168.2.1 on vty 1
　　　　　　　Welcome to DCR Multi-Protocol 1700 Series Router
Router1700>

任务三　交换机端口监听

♂ 需求分析

集线器无论接收到什么数据，都会将数据按照广播的方式在各个端口发送出去，这个方式虽然造成了网络带宽的浪费，但网管设备对网络数据的收集和监听是很有效的；交换机在收到数据帧之后，会根据目的地址的类型决定是否需要转发数据，而且如果不是广播数据，则它只会将数据发送给某一个特定的端口，这样的方式对网络效率的提高很有好处，但对于网管设备来说，在交换机连接的网络中监视所有端口的往来数据似乎变得更困难了。

♂ 方案设计

解决这个问题的办法之一就是在交换机中做配置，使交换机将某一端口的流量在必要的时候镜像给网管设备所在端口，从而实现网管设备对某一端口的监视。这个过程被称为"端口镜像"。

在交换式网络中，对网络数据的分析工作并没有像人们预想的那样变得更加快捷，由于交换机是进行定向转发的设备，因此网络中其他不相关的端口将无法收到其他端口的数据，如网管的协议分析软件安装在一台接在端口 1 中的机器上，而如果想分析端口 2 与端口 3 设备之间的数据流量就变得几乎不可能了。

本任务使用 DCS-3926S 系列交换机作为演示设备，其软件版本为 DCS-3926S_6.1.12.0，实际使用中由于软件版本不同，其功能和配置方法有可能存在差异，请关注相应版本的使用说明。

所需设备如图 1-3-1 所示。

（1）DCS 二层交换机 1 台。
（2）PC 3 台。
（3）Console 线 1 条。
（4）直通网线 3 条。

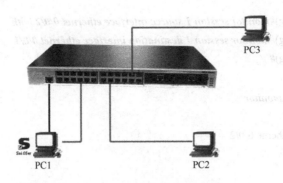

图 1-3-1　交换机端口监听拓扑图

各 PC 网络参数设置见表 1-3-1。

表 1-3-1　各 PC 网络参数设置

设备	IP 地址	子网掩码	端口
PC1	192.168.1.101	255.255.255.0	交换机 e0/0/1
PC2	192.168.1.102	255.255.255.0	交换机 e0/0/2
PC3	192.168.1.103	255.255.255.0	交换机 e0/0/3

知识准备

（1）DCS-3926S 目前只支持一个镜像目的端口，镜像源端口没有使用上的限制，可以有一个，也可以有多个，多个源端口可以处于相同的 VLAN，也可以处于不同的 VLAN。但如果镜像目的端口能镜像到多个镜像源端口的流量，则镜像目的端口必须同时属于这些镜像源端口的所在的 VLAN。

（2）镜像目的端口不能是端口聚合组成员。

（3）镜像目的端口的吞吐量如果小于镜像源端口吞吐量的总和，则目的端口无法完全复制源端口的流量；可减少源端口的个数或复制单向的流量，或者选择吞吐量更大的端口作为目的端口。

端口镜像技术可以将一个源端口的数据流量完全镜像到另一个目的端口进行实时分析。利用端口镜像技术，可以把端口 2 或端口 3 的数据流量完全镜像到端口 1 中进行分析。端口镜像完全不影响镜像端口的工作。

任务实现

步骤1：交换机全部恢复出厂设置后，配置端口镜像，将端口2或者端口3的流量镜像到端口1中。

```
DCS-3926S(Config)#monitor session 1 source interface ethernet 0/0/2 ?
  both              -- Monitor received and transmitted traffic
  rx                -- Monitor received traffic only
  tx                -- Monitor transmitted traffic only
  <CR>
DCS-3926S(Config)#monitor session 1 source interface ethernet 0/0/2 both
DCS-3926S(Config)#monitor session 1 destination interface ethernet 0/0/1
DCS-3926S(Config)#
```

步骤2：验证配置。

```
DCS-3926S#show monitor
session number : 1
Source ports:      Ethernet0/0/2
RX: No
TX: No
Both: Yes
Destination port: Ethernet0/0/1
-------------------------------------------------
DCS-3926S#
```

步骤3：启动抓包软件，PC2 ping PC3，查看是否可以捕捉到数据包，如图1-3-2所示。

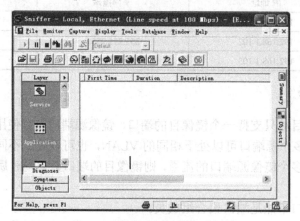

图1-3-2　抓包软件捕捉数据包

任务四　交换机链路聚合

需求分析

两个实验室分别使用一台交换机提供20多个信息点，两个实验室的互连通过一条级联网线实现。每个实验室的信息点都是百兆到桌面,两个实验室之间的带宽也是100Mb/s,如果实验室之间需要大量传输数据,就会明显感觉带宽资源紧张。当楼层之间大量用户都希望以100Mb/s 传输数据的时候,楼层间的链路就呈现出了"独木桥"的状态,必然会造成网络传输效率下降等后果。

♂ 方案设计

解决这个问题的办法就是提高楼层主交换机之间的连接带宽,实现的办法之一是采用千兆端口替换原来的百兆端口进行互连,但这样无疑会增加组网的成本,需要更新端口模块,并且线缆也需要做进一步的升级。而相对经济的升级办法就是链路聚合技术。

本任务使用 DCS-3926S 系列交换机作为演示设备,其软件版本为 DCS-3926S_6.1.12.0,实际使用中由于软件版本不同,功能和配置方法将可能存在差异,请关注相应版本的使用说明。

所需设备如图 1-4-1 所示。
(1) DCS 二层交换机 2 台。
(2) PC 2 台。
(3) Console 线 1 或 2 条。
(4) 直通网线 4~8 条。

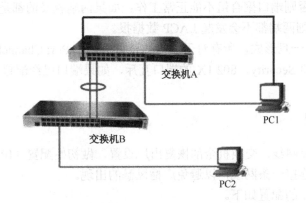

图 1-4-1 交换机链路聚合拓扑图

任务要求见表 1-4-1。

表 1-4-1 各 PC 网络参数设置

设备	IP 地址	子网掩码	端口
交换机 A	192.168.1.11	255.255.255.0	0/0/1-2 聚合
交换机 B	192.168.1.12	255.255.255.0	0/0/3-4 聚合
PC1	192.168.1.101	255.255.255.0	交换机 A0/0/23
PC2	192.168.1.102	255.255.255.0	交换机 B0/0/24

如果链路聚合成功,则 PC1 可以连通 PC2。

♂ 知识准备

链路聚合是指对几个链路做聚合处理,这几个链路必须是同时连接两个相同的设备的,做了链路聚合之后就可以实现几个链路相加的带宽了。例如,可以将 4 个 100Mb/s 的链路使用链路聚合成一个逻辑链路，这样在全双工条件下就可以达到 800Mb/s 的带宽，即将近 1000Mb/s 的带宽。这种方式比较经济，实现也相对容易一些。

（1）为了使 Port Channel 正常工作，Port Channel 的成员端口必须具备以下相同的属性。
① 端口均为全双工模式。
② 端口速率相同。
③ 端口的类型必须一样，如同为以太口或同为光纤口。
④ 端口同为 Access 并且属于同一个 VLAN 或同为 Trunk 端口。
⑤ 如果端口为 Trunk 端口，则其 Allowed VLAN 和 Native VLAN 属性也应该相同。

（2）支持任意两个交换机物理端口的汇聚，最大组数为 6 个，组内最多的端口数为 8 个。

（3）一些命令不能在 Port Channel 的端口上使用，包括 arp、bandwidth、ip、ip-forward 等。

（4）在使用强制生成端口聚合组时，由于汇聚是手工配置触发的，如果由于端口的 VLAN 信息不一致而导致汇聚失败，则汇聚组一直会停留在没有汇聚的状态，必须通过向该 group 增加和删除端口来触发端口再次汇聚，如果 VLAN 信息还是不一致，则仍然不能汇聚成功。直到 VLAN 信息都一致并且有增加和删除端口触发汇聚的情况时，端口才能汇聚成功。

（5）检查对端交换机的对应端口是否配置端口聚合组，且要查看配置方式是否相同，如果本端是手工方式，则对端也应该配置成手工方式，如果本端是 LACP 动态生成，则对端也应该是 LACP 动态生成，否则端口聚合组不能正常工作；如果两端收发的都是 LACP，则至少有一端是 ACTIVE 的，否则两端都不会发起 LACP 数据报。

（6）Port Channel 一旦形成，所有对于端口的设置只能在 Port Channel 端口上进行。

（7）LACP 必须和 Security、802.1X 的端口互斥，如果端口已经配置上述两种协议，则不允许启用 LACP。

任务实现

步骤 1：正确连接网线，交换机全部恢复出厂设置，做初始配置（IP 地址为可选配置）。注意，交换机之间只连接一条网线，以避免广播风暴的出现。

步骤 2：交换机 A 的配置如下。

```
switch#config
switch(Config)#hostname switchA
switchA(Config)#interface vlan 1
switchA(Config-If-Vlan1)#ip address 192.168.1.11 255.255.255.0
switchA(Config-If-Vlan1)#no shutdown
switchA(Config-If-Vlan1)#exit
```

步骤 3：交换机 B 的配置如下。

```
switch#config
switch(Config)#hostname switchB
switchB(Config)#interface vlan 1
switchB(Config-If-Vlan1)#ip address 192.168.1.12 255.255.255.0
switchB(Config-If-Vlan1)#no shutdown
switchB(Config-If-Vlan1)#exit
```

步骤 4：在交换机 A 上创建 port-group。

```
switchA(Config)#port-group 1
```

switchA(Config)#

步骤 5：验证交换机 A 的配置。

switchA#*show port-group detail*
Sorted by the ports in the group 1:
----

switchA#*show port-group brief*
Port-group number : 1
Number of ports in port-group : 0   Maxports in port-channel = 8
Number of port-channels : 0   Max port-channels : 1
switchA#

步骤 6：在交换机 B 上创建 port-group。

switchB(Config)#*port-group 2*
switchB(Config)#

步骤 7：在交换机 A 上手工生成链路聚合通道。

switchA(Config)#*interface ethernet 0/0/1-2*
switchA(Config-Port-Range)#*port-group 1 mode on*   //注意双端模式的匹配性
switchA(Config-Port-Range)#*exit*
switchA(Config)#interface port-channel 1
switchA(Config-If-Port-Channel1)#

步骤 8：在交换机 A 上验证配置。

switchA#*show vlan*

VLAN	Name	Type	Media	Ports	
1	default	Static	ENET	Ethernet0/0/3	Ethernet0/0/4
				Ethernet0/0/5	Ethernet0/0/6
				Ethernet0/0/7	Ethernet0/0/8
				Ethernet0/0/9	Ethernet0/0/10
				Ethernet0/0/11	Ethernet0/0/12
				Ethernet0/0/13	Ethernet0/0/14
				Ethernet0/0/15	Ethernet0/0/16
				Ethernet0/0/17	Ethernet0/0/18
				Ethernet0/0/19	Ethernet0/0/20
				Ethernet0/0/21	Ethernet0/0/22
				Ethernet0/0/23	Ethernet0/0/24
				Port-Channel1	

switchA#   //port-channel1 已经存在

步骤 9：在交换机 B 上手工生成链路聚合通道。

switchB(Config)#*int e 0/0/3-4*
switchB(Config-Port-Range)#*port-group 2 mode on*
switchB(Config-Port-Range)#*exit*
switchB(Config)#*interface port-channel 2*
switchB(Config-If-Port-Channel2)#

步骤 10：在交换机 B 上验证配置。

```
switchB#show port-group brief
Port-group number : 2
Number of ports in port-group : 2    Maxports in port-channel = 8
Number of port-channels : 1          Max port-channels : 1
switchB#
```

步骤 11：使用 ping 命令验证配置，使用 PC1 ping PC2 验证结果是否正确，见表 1-4-2。

表 1-4-2 验证结果表

交换机 A	交换机 B	结果	原因
0/0/1 0/0/2	0/0/3 0/0/4	通	链路聚合组连接正确
0/0/1 0/0/2	0/0/3	通	拔掉交换机 B 端口 4 的网线，仍然可以连通（需要一些时间），此时用 show vlan 查看结果，port-channel 消失。只有一个端口连接的时候，没有必要维持一个 port-channel
0/0/1 0/0/2	0/0/5 0/0/6	通	等候一小段时间后，仍然是连通的。用 show vlan 查看结果。此时，使两台交换机的 spanning-tree 功能 disable，再使用步骤 3 和步骤 4 结果会不同。采用步骤 4 的，将会形成环路

任务五　交换机系统升级与备份

♂ 需求分析

对交换机做好相应的配置后，聪明的管理员会把运行稳定的配置文件和系统文件从交换机中复制出来并保存在稳妥的地方，防止日后因为交换机出现故障而导致配置文件丢失的情况发生。

♂ 方案设计

有了保存的配置文件和系统文件，当交换机被清空之后，可以直接把备份的文件下载到交换机上，避免重新配置的麻烦。交换机文件的备份需要采用 TFTP 服务器（或 FTP 服务器），这也是目前最流行的上传下载的方法。

所需设备如图 1-5-1 所示。

（1）DCS 二层交换机 1 台。
（2）PC 1 台、TFTP Server 1 台（1 台 PC 也可以，既作为调试机又作为 TFTP 服务器）。
（3）Console 线 1 条。
（4）直通网线 1 条。

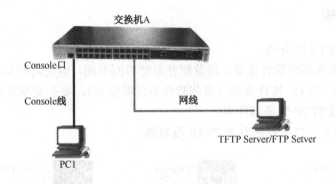

图 1-5-1　交换机系统升级与备份拓扑图

任务要求如下。
（1）按照拓扑图连接网络。
（2）PC 和交换机的 24 口用网线相连。
（3）交换机的管理 IP 地址为 192.168.1.100/24。
（4）PC 网卡的 IP 地址为 192.168.1.101/24。

知识准备

TFTP（Trivial File Transfer Protocol，简单文件传输协议）/FTP（File Transfer Protocol，文件传输协议）都是文件传输协议，在 TCP/IP 协议族中处于第四层，即属于应用层协议，主要用于主机之间、主机与交换机之间的文件传输。它们都采用客户机/服务器模式进行文件传输。

TFTP 承载在 UDP 之上，提供不可靠的数据流传输服务，不提供用户认证机制及根据用户权限对文件操作进行授权的服务；它通过发送包文、应答方式、加上超时重传方式来保证数据的正确传输。TFTP 相对于 FTP 的优点是提供简单的、开销不大的文件传输服务。

FTP 承载在 TCP 之上，提供可靠的面向连接数据流的传输服务，但它不提供文件存取授权，以及简单的认证机制（通过明文传输用户名和密码来实现认证）。FTP 在进行文件传输时，客户机和服务器之间要建立两个连接：控制连接和数据连接。首先由 FTP 客户机发出传送请求，与服务器的 21 号端口建立控制连接，再通过控制连接来协商数据连接。

由此可见，两种方式有其不同的应用环境，局域网内备份和升级可以采用 TFTP 方式，广域网中备份和升级则最好使用 FTP 方式。

如果交换机真的出现了故障，那么会用到本任务的内容：把原来的系统文件和配置文件导入交换机，称为文件还原；把最新的系统文件导入交换机替换原来的系统文件，称为系统升级。神州数码会把每款产品的最新的系统文件放在 www.dcnetworks.com.cn 上免费供用户下载。新的系统文件会修正原版本的一些问题，或者增加一些新功能。对于交换机用户来说，不一定需要时时关注系统文件的最新版本，只要交换机在目前的网络环境中能正常稳定的工作，就不需要升级。

文件的上传下载也是大家经常听到的专业术语。文件上传对应文件备份，文件下载对应系统升级和文件还原。上传和下载是从 TFTP/FTP 服务器的角度来说的，客户机把文件传输给服务器称为上传，客户机从服务器上取得文件称为下载。

任务实现

步骤 1：配置 TFTP 服务器。

市场上 TFTP 服务器的软件很多，每款软件虽然界面不同，但是功能都是一样的，使用方法也都类似：首先是 TFTP 软件安装（有些软件不需要安装），安装完毕之后设定根目录，需要使用的时候，开启 TFTP 服务器即可。

图 1-5-2 是市场上比较流行的几款 TFTP 服务器。

图 1-5-2　三款 TFTP 服务器

步骤 2：这里以第一种 TFTP 服务器为例进行介绍，Tftpd32.exe 非常简单易学，它甚至不需要安装即可使用（后两款软件都需要安装）。双击 Tftpd32.exe，进入如图 1-5-3 所示的 TFTP 服务器的主界面。

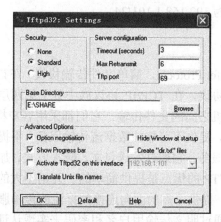

图 1-5-3　TFTP 服务器主界面

步骤 3：在主界面中可以看到该服务器的根目录是 E:\SHARE，服务器的 IP 地址也自动出现在第二行，即 192.168.1.101。可以更改根目录到需要的任何位置，单击"Settings"按钮即可，如图 1-5-4 所示。

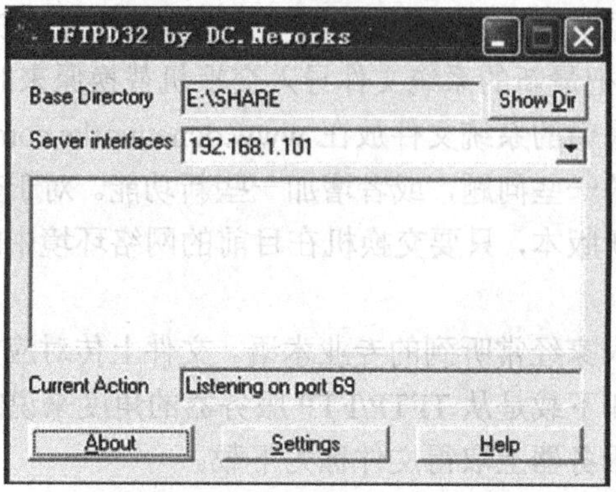

图 1-5-4　设置界面

步骤 4：单击"Browse"按钮进行设置，单击"OK"按钮进行保存。此时，TFTP 服务器就已经配置好了。可以将它最小化到右下角的工具栏中。

步骤 5：设置交换机 IP 地址，即管理 IP 地址。

```
switch(Config)#interface vlan 1                          //进入 VLAN 1 接口
switch(Config-If-Vlan1)#ip address 192.168.1.100 255.255.255.0
switch(Config-If-Vlan1)#no shutdown                      //激活 VLAN 接口
switch(Config-If-Vlan1)#exit
switch(Config)#exit
switch#
```

步骤 6：验证主机与交换机是否连通。

```
switch#ping 192.168.1.101
Type ^c to abort.
Sending 5 56-byte ICMP Echos to 192.168.1.101, timeout is 2 seconds.
!!!!!                             //5 个感叹号表示 5 个包都 ping 通了
Success rate is 100 percent (5/5), round-trip min/avg/max = 1/1/1ms
switch#
```

步骤 7：查看备份的文件。

```
switch#show flash
file name              file length
nos.img                1720035 bytes          //系统文件
startup-config         862 bytes              //该配置文件需要保存
running-config         862 bytes              //该文件和 startup-config 是一样的
switch#
```

步骤 8：备份配置文件。

```
switch#copy startup-config tftp://192.168.1.101/startup1
Confirm [Y/N]:y
begin to send file,wait...
file transfers complete.
close tftp client.
switch#
```

步骤 9：验证是否成功。到 TFTP 服务器根目录中查看文件是否存在，大小是否相同，如图 1-5-5 所示。

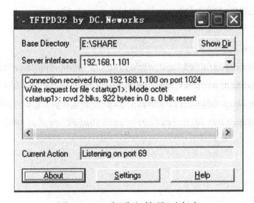

图 1-5-5　查看文件是否存在

步骤10：备份系统文件。

```
DCS-3926S#copy nos.img tftp://192.168.1.101/nos.img
Confirm [Y/N]:y
nos.img file length = 1720035
read file ok
begin to send file,wait...
####################################################################
####################################################################
####################################################################
####################################################################
##############
file transfers complete.
close tftp client.
DCS-3926S#
```

步骤11：对当前的配置做修改并进行保存。

```
switch#config
switch(Config)#hostname DCS-3926S
DCS-3926S(Config)#exit
DCS-3926S#write
DCS-3926S#
```

步骤12：下载配置文件。

```
DCS-3926S#copy tftp://192.168.1.101/startup_20060101 startup-config
Confirm [Y/N]:y
begin to receive file,wait...
recv 865
write ok
transfer complete
close tftp client.
DCS-3926S#
```

步骤13：重新启动并验证是否已经还原。

```
DCS-3926S#reload//重新启动完成之后，标识符是"switch"，表明任务成功
```

步骤14：交换机升级，下载升级包到TFTP服务器中。

```
DCS-3926S#copy tftp://192.168.1.101/nos.img nos.img
Confirm [Y/N]:y
begin to receive file,wait...
####################################################################
####################################################################
####################################################################
####################################################################
################
recv 3330245
begin writing flash.............................
end writing flash.
write ok
```

```
transfer complete
close tftp client.
DCS-3926S#reload
DCS-3926S#show version
```

任务六　路由器系统升级与备份

♂ 需求分析

小张是公司的网络管理员,对路由器做好相应的配置之后,他想把运行稳定的配置文件和系统文件从路由器中复制出来并保存在稳妥的地方,防止日后因为路由器出现故障而导致配置文件丢失的情况发生。

♂ 方案设计

路由器生产厂家会不断推出新的软件版本,增加新的功能,管理员需要及时升级。 配置文件需要及时备份,当文件损坏,或者设备更换时能快速恢复。正常情况下,可以通过 TFTP 或者 FTP 进行升级与备份,当设备无法正常启动时,可以通过 Zmodem 方式恢复。

所需设备如图 1-6-1 所示。

(1) DCR 路由器 1 台。
(2) PC 1 台。
(3) TFTP 软件。

图 1-6-1　路由器系统升级与备份

任务要求:配置设备的 F0/0 与网卡 IP 地址分别为 192.168.2.1 和 192.168.2.10,通过 TFTP 软件对路由器的操作系统进行升级与备份。

♂ 任务实现

1. 通过 TFTP 方式进行备份

步骤 1:设置 PC 网卡 IP 地址为 192.168.2.10,并安装 TFTP Server,此处略。
步骤 2:参照相关实验,设置路由器的 F0/0 接口 IP 地址为 192.168.2.1,并测试连通性。

```
Router-A#config
Router-A_config#interface f0/0
Router-A_config_f0/0#ip address 192.168.2.1 255.255.255.0
Router-A_config_f0/0#no shutdown
Router-A_config_f0/0#^Z
Router-A#show interface f0/0
FastEthernet0/0 is up, line protocol is up
```

```
    address is 00e0.0f18.1a70
      Interface address is 192.168.2.1/24
      MTU 1500 bytes, BW 100000 kbit, DLY 10 usec
      Encapsulation ARPA, loopback not set
   Keepalive not set
   <省略…>
   Router-A#ping 192.168.2.10                    //ping PC 的地址
   PING 192.168.2.10 (192.168.2.1): 56 data bytes
   !!!!!
   --- 192.168.2.1 ping statistics ---
   5 packets transmitted, 5 packets received, 0% packet loss
   round-trip min/avg/max = 20/22/30 ms
```

步骤 3：查看路由器文件，并将配置文件下载到 TFTP 服务器中。

```
   Router-A#dir
   Directory of /:
   2     DCR-1751.bin         <FILE>     3589526    Sun Feb  7 06:28:15 2106
   3     startup-config<FILE>            516        Thu Jan  1 00:03:09 2004
   free space 4751360
   Router-A#copyflash:startup-configtftp:          //上传配置文件作为备份
   Remote-server ip address[]?192.168.2.10         //TFTP 服务器的 IP 地址
   Destination file name[startup-config]?          //默认的文件名
   #
   TFTP:successfully send 2 blocks ,516 bytes
```

步骤 4：使用写字板打开下载后的配置文件，修改机器名，再次使用 copy 命令上传到路由器中，重新启动后通过 show 命令查看到机器名已经被修改。

2. 使用 TFTP 维护路由器的原系统备份并升级成新版本

步骤 1：使用 show version 命令查看当前系统版本。

```
   Router#show version
   Digitalchina Internetwork Operating System Software
   1700 Series Software, Version 1.3.2E (BASE), RELEASE SOFTWARE
   Copyright (c) 1996-2000 by ChinaDigitalchina CO.LTD
   Compiled: 2005-4-8 9:12:34 by system, Image text-base: 0x6004
   ROM: System Bootstrap, Version 0.1.8
   Serial num:8IRT01V11802000044, ID num:000044
   System image file is "base.bin"
   DCR-1700 Processor MPC860T CPU at 50Mhz
   32768K bytes of memory,8192K bytes of flash
   Router uptime is 0:00:07:50, The current time: 2004-1-1 0:7:50
   Slot 0: FEC Slot
      Port 0: 10/100Mbps full-duplex Ethernet
   Slot 1: SCC Slot
      Port 0: 10M Ethernet
      Port 1: 10M Ethernet
   Slot 2: SCC Slot
      Port 0: serial
```

Router#

步骤 2：使用 dir 命令查看当前系统文件名称。

Router#dir
Directory of /:
0   base.bin<FILE>   3589526   Sun Nov 2 21:56:51 1980
1   startup-config<FILE>   786   Thu Jan 1 01:02:32 2004
free space 4751360
Router#

步骤 3：确保 TFTP 服务器与路由器连通。

Router#ping 192.168.2.10
PING 192.168.2.10 (192.168.2.10): 56 data bytes
!!!!!
--- 192.168.2.10 ping statistics ---
5 packets transmitted, 5 packets received, 0% packet loss
round-trip min/avg/max = 0/0/0 ms
Router#

步骤 4：开启 TFTP 服务器，使用 copy file tftp 命令备份系统文件，如图 1-6-2 所示。

Router#copy flash: tftp:
Source file name[]?base.bin
Remote-server ip address[]?192.168.2.10
Destination file name[base.bin]?132obase.bin
##
##
##
##
##
##
##
##
##
##
##
TFTP:successfully send 7011 blocks ,3589526 bytes
Router#

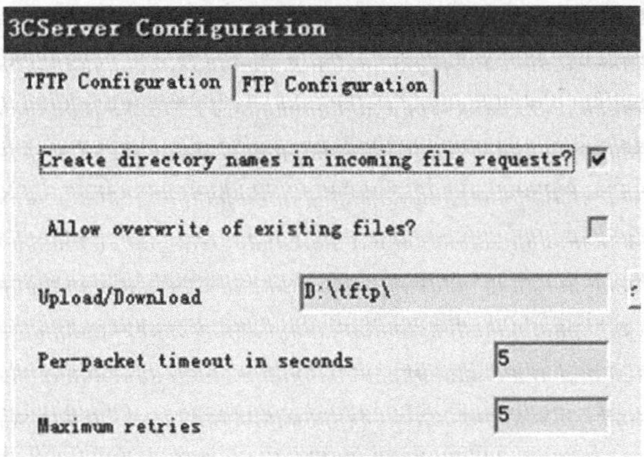

图 1-6-2 开启 TFTP 服务器并设置目录

步骤 5：在 TFTP 服务器中查看保存的文件，打开 TFTP 服务器的主目录，可以看到文件，如图 1-6-3 所示。

图 1-6-3　在 TFTP 服务器主目录中查看文件

步骤 6：在神州数码网站下载路由器新版本到 TFTP 服务器主目录中，下载地址是 http://www.dcnetworks.com.cn/cn/download/，找到合适的设备型号开始下载。查看当前 TFTP 服务器的上传下载主目录，如图 1-6-4 所示。

图 1-6-4　查看 TFTP 服务器的上传下载主目录

步骤 7：使用 copy tftp file 命令升级新版本。

```
Router#copytftp: flash:
Source file name[]?DCR1700-1.3.3A-M.bin
Remote-server ip address[]?192.168.2.10
Destination file name[DCR1700-1.3.3A-M.bin]?
################################################################
################################################################
################################################################
################################################################
################################################################
################################################################
################################################################
################################################################
################################################################
################################################################
################################################################
################################################################
################################################################
```

```
################################################################
################################################################
##############################################TFTP:file close error
TFTP:successfully receive 9340 blocks ,4782079 bytes
Router#
```

认证考核

一、选择题

1. 交换机或路由器的网络操作系统存储在（　　）存储器中。
   A．ROM　　　　B．NVROM　　　　C．FLASH　　　　D．DRAM
2. 交换机或路由器的配置文件保存在（　　）存储器中。
   A．ROM　　　　B．NVROM　　　　C．FLASH　　　　D．DRAM
3. 若要查看交换机的当前配置，以下命令中，正确的是（　　）。

   A．switch>show run　　　　　　　B．<switch>show run
   C．<switch>disp cur　　　　　　　D．switch#show run
4. 若要设置交换机主机名为student1，以下配置命令中，正确的是（　　）。

   A．switch>sysname student1　　　　B．switch#sysname student1
   C．switch(Config)#hostname student1　　D．switch#hostname student1
5. 新购买回来的交换机进行首次配置后，应采用的配置方式是（　　）。

   A．通过以太网口，利用超级终端进行配置
   B．通过 Console 口，利用超级终端进行配置
   C．通过以太网口，利用 Telnet 进行配置
   D．通过 Console 口，利用 Web 界面进行配置
6. Console 口默认的通信波特率为（　　）。

   A．4800b/s　　　B．9600b/s　　　C．115200b/s　　　D．2400b/s
7. Telnet 使用的端口号是（　　）。

   A．TCP 22　　　B．TCP 23　　　C．UDP 22　　　D．UDP 23

二、操作题

1. 设置特权用户配置模式的 enable 明文形式密码为"digitalchina"。
2. 设置交换机的时间为当前时间。
3. 设置交换机名称为 digitalchina-3950S。
4. 请把交换机的帮助信息设置为中文。
5. 交换机恢复出厂设置。
6. 使用各种 TFTP 软件进行 TFTP 或者 FTP 的文件备份。

项目二 VLAN 技术与生成树技术

教学背景

VLAN（Virtual Local Area Network，虚拟局域网）是在一个物理网络上划分出来的逻辑网络。这个网络对应于 OSI 模型的第二层。通过将企业网络划分为 VLAN，可以强化网络管理和网络安全，控制不必要的数据广播。VLAN 将网络划分为多个广播域，从而有效地控制广播风暴的发生，还可以用于控制网络中不同部门、不同站点之间的互相访问。

人们对网络的依赖性越来越强，为了保证网络的高可用性，有时希望在网络中提供设备、模块和链路的冗余。但是在二层网络中，冗余链路可能会导致交换环路，使得广播包在交换环路中无休止地循环，进而破坏网络中设备的工作性能，甚至导致整个网络瘫痪。生成树技术能够解决交换环路的问题，并为网络提供冗余。

任务一 实现跨交换机相同 VLAN 内通信

需求分析

某教学楼有两层，分别是一年级、二年级，每个楼层都有一台交换机满足教师上网需求；每个年级都有语文教研组和数学教研组；两个年级的语文教研组的计算机可以互相访问；两个年级的数学教研组的计算机可以互相访问；语文教研组和数学教研组之间不可以自由访问。

方案设计

通过划分 VLAN 可以实现跨交换机的相同 VLAN 通信，使得语文教研组和数学教研组之间不可以自由访问；使用 802.1Q 可以进行跨交换机的 VLAN。

所需设备如图 2-1-1 所示。

（1）DCS 二层交换机 2 台。
（2）PC 2 台。
（3）Console 线 1 条。
（4）直通网线 2 条。

项目二

VLAN 技术与生成树技术

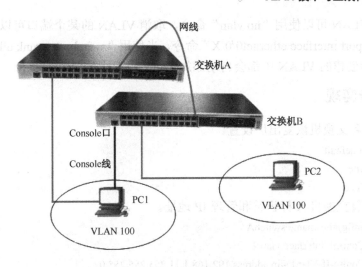

图 2-1-1 跨交换机相同 VLAN 通信拓扑图

任务要求：在交换机 A 和交换机 B 上分别划分两个基于端口的 VLAN——VLAN100、VLAN200。使得交换机之间 VLAN100 的成员能够互相访问，VLAN200 的成员能够互相访问；VLAN100 和 VLAN200 成员之间不能互相访问，见表 2-1-1 和表 2-1-2。

表 2-1-1 划分 VLAN

VLAN	端口成员
100	1~8
200	9~16
Trunk 口	24

表 2-1-2 PC1 和 PC2 的网络设置

设备	IP 地址	子网掩码
交换机 A	192.168.1.11	255.255.255.0
交换机 B	192.168.1.12	255.255.255.0
PC1	192.168.1.101	255.255.255.0
PC2	192.168.1.102	255.255.255.0

PC1、PC2 分别接在不同交换机 VLAN100 的成员端口 1~8 上，两台 PC 互相可以 ping 通；PC1、PC2 分别接在不同交换机 VLAN 的成员端口 9~16 上，两台 PC 互相可以 ping 通；PC1 和 PC2 接在不同 VLAN 的成员端口上，互相 ping 不通。

若实验结果和理论相符，则本任务完成。

知识准备

二层交换机只有二层的功能，用于实现一个广播域内的主机间的通信。在 OSI 模型中，这个设备工作在数据链路层。三层交换机就是在二层的功能上添加一个路由模块，实现三层的路由转发功能，即三层交换机可以同时工作在数据链路层和网络层上。二层交换机同一 VLAN 中的端口能互相通信，不同 VLAN 中的端口不能互相通信。二层交换机不具备路由功能。

取消一个 VLAN 可以使用"no vlan"命令，取消 VLAN 的某个端口可以在 VLAN 模式下使用"no switchport interface ethernet0/0/X"命令，当使用"switchport trunk allowed vlan all"命令后，所有以后创建的 VLAN 中都会自动添加 Trunk 口为成员端口。

任务实现

步骤1：将各交换机恢复出厂设置。

```
switch#set default
switch#write
switch#reload
```

步骤2：设置交换机 A 标识符和管理 IP 地址。

```
switch(Config)#hostname switchA
switchA(Config)#interface vlan 1
switchA(Config-If-Vlan1)#ip address 192.168.1.11 255.255.255.0
switchA(Config-If-Vlan1)#no shutdown
switchA(Config-If-Vlan1)#exit
switchA(Config)#
```

步骤3：设置交换机 B 标识符和管理 IP 地址。

```
switch(Config)#hostname switchB
switchB(Config)#interface vlan 1
switchB(Config-If-Vlan1)#ip address 192.168.1.12 255.255.255.0
switchB(Config-If-Vlan1)#no shutdown
switchB(Config-If-Vlan1)#exit
switchB(Config)#
```

步骤4：在交换机 A 中创建 VLAN100 和 VLAN200，并添加端口。

```
switchA(Config)#vlan 100
switchA(Config-Vlan100)#
switchA(Config-Vlan100)#switchport interface ethernet 0/0/1-8
switchA(Config-Vlan100)#exit
switchA(Config)#vlan 200
switchA(Config-Vlan200)#switchport interface ethernet 0/0/9-16
switchA(Config-Vlan200)#exit
switchA(Config)#
```

步骤5：验证交换机 A 的 VLAN 配置。

```
switchA#show vlan
VLAN Name         Type      Media    Ports
-----------------------------------------------------------
1    default      Static    ENET     Ethernet0/0/17    Ethernet0/0/18
                                     Ethernet0/0/19    Ethernet0/0/20
                                     Ethernet0/0/21    Ethernet0/0/22
     Ethernet0/0/23    Ethernet0/0/24
100  VLAN0100     Static    ENET     Ethernet0/0/1     Ethernet0/0/2
                                     Ethernet0/0/3     Ethernet0/0/4
                                     Ethernet0/0/5     Ethernet0/0/6
```

项目二 VLAN 技术与生成树技术

				Ethernet0/0/7	Ethernet0/0/8
200	VLAN0200	Static	ENET	Ethernet0/0/9	Ethernet0/0/10
				Ethernet0/0/11	Ethernet0/0/12
				Ethernet0/0/13	Ethernet0/0/14
				Ethernet0/0/15	Ethernet0/0/16

switchA#

步骤 6：交换机 B 上的配置与交换机 A 一样，此处略。

步骤 7：设置交换机 A 的 Trunk 端口。

switchA(Config)#interface ethernet 0/0/24
switchA(Config-Ethernet0/0/24)#switchport mode trunk //设置为 Trunk 模式
Set the port Ethernet0/0/24 mode TRUNK successfully
switchA(Config-Ethernet0/0/24)#switchport trunk allowed vlan all
//允许所有 VLAN 通过
set the port Ethernet0/0/24 allowed vlan successfully
switchA(Config-Ethernet0/0/24)#exit
switchA(Config)#
switchA#show vlan

VLAN	Name	Type	Media	Ports	
1	default	Static	ENET	Ethernet0/0/17	Ethernet0/0/18
				Ethernet0/0/19	Ethernet0/0/20
				Ethernet0/0/21	Ethernet0/0/22
				Ethernet0/0/23	
				Ethernet0/0/24(T)	
100	VLAN0100	Static	ENET	Ethernet0/0/1	Ethernet0/0/2
				Ethernet0/0/3	Ethernet0/0/4
				Ethernet0/0/5	Ethernet0/0/6
				Ethernet0/0/7	Ethernet0/0/8
				Ethernet0/0/24(T)	
200	VLAN0200	Static	ENET	Ethernet0/0/9	Ethernet0/0/10
				Ethernet0/0/11	Ethernet0/0/12
				Ethernet0/0/13	Ethernet0/0/14
				Ethernet0/0/15	Ethernet0/0/16
				Ethernet0/0/24(T)	

switchA#

24 口已经出现在 VLAN1、VLAN100 和 VLAN200 中，并且 24 口不是一个普通端口，是 tagged 端口。

步骤 8：设置交换机 B 的 Trunk 端口同交换机 A 一样，此处略。

步骤 9：验证。交换机 A ping 交换机 B。

switchA#ping 192.168.1.12
Type ^c to abort.
Sending 5 56-byte ICMP Echos to 192.168.1.12, timeout is 2 seconds.
!!!!!
Success rate is 100 percent (5/5), round-trip min/avg/max = 1/1/1ms

switchA#

这表明交换机之间的 Trunk 链路已经成功建立。

按表 2-1-3 验证，PC1 安装在交换机 A 上，PC2 安装在交换机 B 上。

表 2-1-3 验证结果

PC1 位置	PC2 位置	动作	结果
1～8 端口		PC1 ping 交换机 B	不通
9～16 端口		PC1 ping 交换机 B	不通
17～24 端口		PC1 ping 交换机 B	通
1～8 端口	1～8 端口	PC1 ping PC2	通
1～8 端口	9～16 端口	PC1 ping PC2	不通

任务二 实现不同 VLAN 间通信

需求分析

某学校软件实验室的 IP 地址段是 192.168.10.0/24，多媒体实验室的 IP 地址段是 192.168.20.0/24，为了保证它们之间的数据互不干扰，也不影响各自的通信效率，网络管理员划分了 VLAN，使两个实验室属于不同的 VLAN。

方案设计

两个实验室有时候也需要相互通信，此时就要利用三层交换机划分 VLAN。

所需设备如图 2-2-1 所示。

（1）DCRS-5650 交换机 1 台（软件版本为 DCRS-5650-28_5.2.1.0）。

（2）PC 2 台。

（3）Console 线 1 条。

（4）直通网线若干。

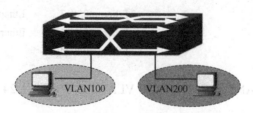

图 2-2-1 实现不同 VLAN 间通信

任务要求：使用一台交换机和两台 PC，将其中的 PC2 作为控制台终端，使用 Console 口配置方式；使用两条网线分别将 PC1 和 PC2 连接到交换机的 RJ-45 接口上。在交换机上划分两个基于端口的 VLAN——VLAN100、VLAN200，见表 2-2-1。

表 2-2-1　在交换机上划分 VLAN

VLAN	端口成员
100	0/0/1～0/0/12
200	0/0/13～0/0/24

使得 VLAN100 的成员能够互相访问，VLAN200 的成员能够互相访问；VLAN100 和 VLAN200 成员之间不能互相访问，见表 2-2-2 和表 2-2-3。

表 2-2-2　PC 配置表（1）

设备	端口	IP 地址	网关 1	子网掩码
交换机		192.168.1.1	无	255.255.255.0
VLAN100		无	无	255.255.255.0
VLAN200		无	无	255.255.255.0
PC1	1～12	192.168.1.101	无	255.255.255.0
PC2	13～24	192.168.1.102	无	255.255.255.0

表 2-2-3　PC 配置表（2）

设备	端口	IP 地址	网关 1	子网掩码
交换机		192.168.1.1		255.255.255.0
VLAN100		192.168.10.1	无	255.255.255.0
VLAN200		192.168.20.1	无	255.255.255.0
PC1	1～12	192.168.10.11	192.168.10.1	255.255.255.0
PC2	13～24	192.168.20.11	192.168.20.1	255.255.255.0

各设备先使用表 2-2-2 的 IP 地址，使 PC1 ping PC2，应该不连通；再按照表 2-2-3 配置各设备的 IP 地址，并在交换机上配置 VLAN 接口的 IP 地址，使 PC1 ping PC2，可连通，该通信属于 VLAN 间通信，要经过三层设备的路由。

若实验结果和理论相符，则本任务完成。

知识准备

与二层交换机不同，三层交换机可以在多个 VLAN 接口上配置 IP 地址。

三层交换机与路由器的区别：三层交换机也具有"路由"功能，它与传统路由器的路由功能总体上是一致的。尽管如此，三层交换机与路由器还是存在着相当大的本质区别的，具体表现在以下几方面。

1. 主要功能不同

虽然三层交换机与路由器都具有路由功能，但不能因此而把它们等同起来，正如现在许多网络设备同时具备多种传统网络设备功能一样，就如现在许多宽带路由器不仅具有路由功能，还提供了交换机端口、硬件防火墙功能，但不能把它与交换机或者防火墙等同起来一样。因为这些路由器的主要功能还是路由功能，其他功能只是其附加功能，其目的是使设备适用面更广、更加实用。这里的三层交换机也一样，它仍是交换机产品，但它具备了一些基本路由功能的交

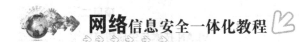

换机，它的主要功能仍是数据交换。也就是说，它同时具备了数据交换和路由转发两种功能，但其主要功能还是数据交换；而路由器仅具有路由转发这一种主要功能。

2. 主要适用的环境不同

三层交换机的路由功能通常比较简单，因为它所面对的主要是简单的局域网连接。正因如此，三层交换机的路由功能通常比较简单，路由路径远没有路由器那么复杂。它用在局域网中的主要用途仍然是提供快速数据交换功能，满足局域网数据交换频繁的应用特点。

而路由器则不同，它的设计初衷就是满足不同类型的网络连接，虽然也适用于局域网之间的连接，但它的路由功能更多地体现在不同类型网络之间的互连上，如局域网与广域网之间的连接、不同协议的网络之间的连接等，所以路由器主要用于不同类型的网络之间。它最主要的功能就是路由转发，解决好各种复杂路由路径网络的连接就是它的最终目的，所以路由器的路由功能通常非常强大，不仅适用于同种协议的局域网之间，更适用于不同协议的局域网与广域网之间。它的优势在于选择最佳路由、负荷分担、链路备份及与其他网络进行路由信息的交换等。为了与各种类型的网络连接，路由器的接口类型非常丰富，而三层交换机一般仅提供同类型的局域网接口，非常简单。

3. 性能体现不同

从技术上讲，路由器和三层交换机在数据包交换操作上存在着明显区别。路由器一般由基于微处理器的软件路由引擎执行数据包交换，而三层交换机通过硬件执行数据包交换。三层交换机在对第一个数据流进行路由后，将会产生一个MAC地址与IP地址的映射表，当同样的数据流再次通过时，将根据此表直接从二层通过而不是再次路由，从而消除了路由器进行路由选择而造成的网络延迟，提高了数据包转发的效率。同时，三层交换机的路由查找是针对数据流的，它利用缓存技术，很容易利用ASIC技术来实现，因此，可以大大节约成本，并实现快速转发。而路由器的转发采用最长匹配的方式，实现复杂，通常使用软件来实现，转发效率较低。

正因如此，从整体性能上比较时，三层交换机的性能要远优于路由器，非常适用于数据交换频繁的局域网中；而路由器虽然路由功能非常强大，但它的数据包转发效率远低于三层交换机，更适用于数据交换不是很频繁的不同类型网络的互连，如局域网与互联网的互连。如果把路由器，特别是高档路由器用于局域网中，则在相当大的程度上是一种浪费（就其强大的路由功能而言），而且不能很好地满足局域网通信性能的需求，影响子网间的正常通信。

♂ 任务实现

步骤1：交换机恢复出厂设置。

```
switch#set default
switch#write
switch#reload
```

步骤2：给交换机设置IP地址，即管理IP地址。

```
switch#config
switch(Config)#interface vlan 1
switch(Config-If-Vlan1)#ip address 192.168.1.1 255.255.255.0
switch(Config-If-Vlan1)#no shutdown
switch(Config-If-Vlan1)#exit
```

```
switch(Config)#exit
```

步骤 3：创建 VLAN100 和 VLAN200。

```
switch(Config)#
switch(Config)#vlan 100
switch(Config-Vlan100)#exit
switch(Config)#vlan 200
switch(Config-Vlan200)#exit
switch(Config)#
```

步骤 4：查看 VLAN。

```
switch#show vlan
VLAN Name           Type      Media   Ports
---- ------------   --------- ------- -----------------------------------------
1    default        Static    ENET    Ethernet0/0/1       Ethernet0/0/2
                                      Ethernet0/0/3       Ethernet0/0/4
                                      Ethernet0/0/5       Ethernet0/0/6
                                      Ethernet0/0/7       Ethernet0/0/8
                                      Ethernet0/0/9       Ethernet0/0/10
                                      Ethernet0/0/11      Ethernet0/0/12
                                      Ethernet0/0/13      Ethernet0/0/14
                                      Ethernet0/0/15      Ethernet0/0/16
                                      Ethernet0/0/17      Ethernet0/0/18
                                      Ethernet0/0/19      Ethernet0/0/20
                                      Ethernet0/0/21      Ethernet0/0/22
                                      Ethernet0/0/23      Ethernet0/0/24
                                      Ethernet0/0/25      Ethernet0/0/26
                                      Ethernet0/0/27      Ethernet0/0/28
100  VLAN0100       Static    ENET
200  VLAN0200       Static    ENET
```

步骤 5：给 VLAN100 和 VLAN200 添加端口。

```
switch(Config)#vlan 100                              ！进入VLAN100
switch(Config-Vlan100)#switchport interface ethernet 0/0/1-12
Set the port Ethernet0/0/1 access vlan 100 successfully
Set the port Ethernet0/0/2 access vlan 100 successfully
Set the port Ethernet0/0/3 access vlan 100 successfully
Set the port Ethernet0/0/4 access vlan 100 successfully
Set the port Ethernet0/0/5 access vlan 100 successfully
Set the port Ethernet0/0/6 access vlan 100 successfully
Set the port Ethernet0/0/7 access vlan 100 successfully
Set the port Ethernet0/0/8 access vlan 100 successfully
Set the port Ethernet0/0/9 access vlan 100 successfully
Set the port Ethernet0/0/10 access vlan 100 successfully
Set the port Ethernet0/0/11 access vlan 100 successfully
Set the port Ethernet0/0/12 access vlan 100 successfully
switch(Config-Vlan100)#exit
```

```
switch(Config)#vlan 200                    ！进入 VLAN200
switch(Config-Vlan200)#switchport interface ethernet 0/0/13-24
Set the port Ethernet0/0/13 access vlan 200 successfully
Set the port Ethernet0/0/14 access vlan 200 successfully
Set the port Ethernet0/0/15 access vlan 200 successfully
Set the port Ethernet0/0/16 access vlan 200 successfully
Set the port Ethernet0/0/17 access vlan 200 successfully
Set the port Ethernet0/0/18 access vlan 200 successfully
Set the port Ethernet0/0/19 access vlan 200 successfully
Set the port Ethernet0/0/20 access vlan 200 successfully
Set the port Ethernet0/0/21 access vlan 200 successfully
Set the port Ethernet0/0/22 access vlan 200 successfully
Set the port Ethernet0/0/23 access vlan 200 successfully
Set the port Ethernet0/0/24 access vlan 200 successfully
switch(Config-Vlan200)#exit
```

步骤 6：查看端口是否接入 VLAN。

```
switch#show vlan
VLAN Name           Type      Media     Ports
---- ------------   --------  --------  ---------------------------------------
1    default        Static    ENET      Ethernet0/0/25      Ethernet0/0/26
                                        Ethernet0/0/27      Ethernet0/0/28
100  VLAN0100       Static    ENET      Ethernet0/0/1       Ethernet0/0/2
                                        Ethernet0/0/3       Ethernet0/0/4
                                        Ethernet0/0/5       Ethernet0/0/6
                                        Ethernet0/0/7       Ethernet0/0/8
                                        Ethernet0/0/9       Ethernet0/0/10
                                        Ethernet0/0/11      Ethernet0/0/12
200  VLAN0200       Static    ENET      Ethernet0/0/13      Ethernet0/0/14
                                        Ethernet0/0/15      Ethernet0/0/16
                                        Ethernet0/0/17      Ethernet0/0/18
                                        Ethernet0/0/19      Ethernet0/0/20
                                        Ethernet0/0/21      Ethernet0/0/22
                                        Ethernet0/0/23      Ethernet0/0/24
switch#
```

步骤 7：验证实验，见表 2-2-4。

表 2-2-4 验证

PC1 位置	PC2 位置	动作	结果
0/0/1~0/0/12 端口	0/0/13~0/0/24 端口	PC1 ping PC2	不通

步骤 8：添加 VLAN 地址。

```
switch(Config)#interface vlan 100
switch(Config-If-Vlan100)#%Jan 01 00:00:59 2006 %LINK-5-CHANGED: Interface Vlan100, changed state to UP
```

switch(Config-If-Vlan100)#ip address 192.168.10.1 255.255.255.0
switch(Config-If-Vlan100)#no shut
switch(Config-If-Vlan100)#exit
switch(Config)#interface vlan 200
switch(Config-If-Vlan200)#%Jan 01 00:00:59 2006 %LINK-5-CHANGED: Interface Vlan100, changed state to UP
switch(Config-If-Vlan200)#ip address 192.168.20.1 255.255.255.0
switch(Config-If-Vlan200)#no shut
switch(Config-If-Vlan200)#exit
switch(Config)#

步骤 9：验证配置。

switch#show ip route
Codes: K - kernel, C - connected, S - static, R - RIP, B - BGP
      O - OSPF, IA - OSPF inter area
      N1 - OSPF NSSA external type 1, N2 - OSPF NSSA external type 2
      E1 - OSPF external type 1, E2 - OSPF external type 2
      i - IS-IS, L1 - IS-IS level-1, L2 - IS-IS level-2, ia - IS-IS inter area
      * - candidate default
C   127.0.0.0/8 is directly connected, Loopback
C   192.168.10.0/24 is directly connected, Vlan100
C   192.168.20.0/24 is directly connected, Vlan200
switch#

步骤 10：验证实验，见表 2-2-5。

表 2-2-5 验证

PC1 位置	PC2 位置	动作	结果
0/0/1～0/0/12 端口	0/0/13～0/0/24 端口	PC1 ping PC2	通

任务三　单实例生成树

♂ 需求分析

交换机之间具有冗余链路本来是一件很好的事情，但是它有可能引起的问题比它能够解决的问题更多。如果真的准备两条以上的路，就必然会形成一条环路，交换机并不知道如何处理环路，只是周而复始地转发帧，形成一个"死循环"，这个死循环会使整个网络处于阻塞状态，导致网络瘫痪。

♂ 方案设计

采用生成树协议可以避免环路。

生成树协议的根本目的是将一个存在物理环路的交换网络变成一个没有环路的逻辑树形网络。IEEE 802.1d 协议通过在交换机上运行一套复杂的算法 STA（Spanning Tree Algorithm，生成树算法），使冗余端口置于"阻断状态"，使得接入网络的计算机在与其他计算机通信时，只有一条链路生效，而当这个链路出现故障无法使用时，IEEE 802.1d 协议会重新计算网络链路，将处于"阻断状态"的端口重新打开，从而既保障了网络正常运转，又保证了冗余能力。

所需设备如图 2-3-1 所示。

（1）DCRS-5650 交换机 2 台（软件版本为 DCRS-5650-28_5.2.1.0）。
（2）PC 2 台。
（3）Console 线 1 或 2 条。
（4）直通网线 4～8 条。

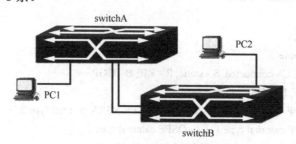

图 2-3-1　单实例生成树

任务要求：IP 地址设置见表 2-3-1，网线连接见表 2-3-2。

表 2-3-1　IP 地址设置

设备	IP 地址	子网掩码
switch A	192.168.1.11	255.255.255.0
switch B	192.168.1.12	255.255.255.0
PC1	192.168.1.101	255.255.255.0
PC2	192.168.1.102	255.255.255.0

表 2-3-2　网线连接

设备 A	设备 B
switch A　e0/0/1	switch B　e0/0/3
switch A　e0/0/2	switch B　e0/0/4
PC1	switch A　e0/0/24
PC2	switch B　e0/0/23

如果生成树成功，则 PC1 可以 ping 通 PC2。

知识准备

（1）如果想在交换机上运行 MSTP，首先必须在全局模式下打开 MSTP 开关。在没有打开全局 MSTP 开关之前，打开端口的 MSTP 开关是不允许的。

（2）MSTP 定时器参数之间是有相关性的，错误配置可能导致交换机不能正常工作。各定时器之间的关联关系为：

$2 \times (Bridge\_Forward\_Delay - 1.0 \text{ seconds}) \geq Bridge\_Max\_Age$

$Bridge\_Max\_Age \geq 2 \times (Bridge\_Hello\_Time + 1.0 \text{ seconds})$

项目二 VLAN 技术与生成树技术

（3）用户在修改 MSTP 参数时，应该清楚所产生的各个拓扑。除了全局的基于网桥的参数配置外，其他的是基于各个实例的配置，在配置时一定要注意配置参数对应的实例是否正确。

（4）DCRS-5650 交换机的端口 MSTP 功能与端口 MAC 绑定、802.1x 和设置端口为路由端口功能互斥。当端口已经配置 MAC 地址绑定、802.1x 或设置为路由端口时，无法在该端口启动 MSTP 功能。

任务实现

步骤 1：正确连接网线，恢复出厂设置之后，做初始配置。

步骤 2：对交换机 A 进行配置。

```
switch#config
switch(Config)#hostname switchA
switchA(Config)#interface vlan 1
switchA(Config-If-Vlan1)#ip address 192.168.1.11 255.255.255.0
switchA(Config-If-Vlan1)#no shutdown
switchA(Config-If-Vlan1)#exit
switchA(Config)#
```

步骤 3：对交换机 B 进行配置。

```
switch#config
switch(Config)#hostname switchB
switchB(Config)#interface vlan 1
switchB(Config-If-Vlan1)#ip address 192.168.1.12 255.255.255.0
switchB(Config-If-Vlan1)#no shutdown
switchB(Config-If-Vlan1)#exit
switchB(Config)#
```

步骤 4：运行"PC1 ping PC2–t"命令，观察现象。

① ping 不通；

② 所有连接网线的端口的绿灯很频繁地闪烁，表明该端口收发数据量很大，已经在交换机内部形成广播风暴。

③ 使用命令"show cpu usage"观察两台交换机 CPU 的使用率。

```
switchA#sh cpu usage
Last    5 second CPU IDLE:    96%
Last 30 second CPU IDLE:    96%
Last    5 minute CPU IDLE:    97%
From    running    CPU IDLE:    97%
switchB#shcpu usage
Last    5 second CPU IDLE:    96%
Last 30 second CPU IDLE:    97%
Last    5 minute CPU IDLE:    97%
From    running    CPU IDLE:    97%
```

步骤 5：在两台交换机中都启用生成树协议。

```
switchA(Config)#spanning-tree
MSTP is starting now, please wait..........
```

MSTP is enabled successfully.
switchA(Config)#
switchB(Config)#spanning-tree
MSTP is starting now, please wait...........
MSTP is enabled successfully.
switchB(Config)#

步骤 6：在交换机 A 上查看配置。

switchA#show spanning-tree
　　　　　　　-- MSTPBridgeConfig Info --
Standard　　　：IEEE 802.1s
Bridge MAC　　：00:03:0f:0f:6e:ad
Bridge Times：　Max Age 20, Hello Time 2, Forward Delay 15
Force Version：　3
######################### Instance 0 #########################
SelfBridge Id　　：32768 -　00:03:0f:0f:6e:ad
Root Id　　　　：32768.00:03:0f:0b:f8:12
Ext.RootPathCost：200000
Region Root Id　　：this switch
Int.RootPathCost：0
Root Port ID　　　：128.1
Current port list in Instance 0:
Ethernet0/0/1 Ethernet0/0/2 (Total 2)
PortName　　　ID　　　ExtRPCIntRPC　State Role　　DsgBridgeDsgPort
--------------- ------- --------- --------- --- ---- ------------------ -------
 Ethernet0/0/1 128.001　　　　0　　　　0 FWD ROOT 32768.00030f0bf812 128.003
 Ethernet0/0/2 128.002　　　　0　　　　0 BLK ALTR 32768.00030f0bf812 128.004

步骤 7：在交换机 B 上查看配置。

switchB#show spanning-tree
　　　　　　　-- MSTPBridgeConfig Info --
Standard　　　：IEEE 802.1s
Bridge MAC　　：00:03:0f:0b:f8:12
Bridge Times：　Max Age 20, Hello Time 2, Forward Delay 15
Force Version：　3
######################### Instance 0 #########################
SelfBridge Id　　：32768 -　00:03:0f:0b:f8:12
Root Id　　　　　：this switch
Ext.RootPathCost：0
Region Root Id　　：this switch
Int.RootPathCost：0
Root Port ID　　　：0
Current port list in Instance 0:
Ethernet0/0/3 Ethernet0/0/4 (Total 2)
PortName　　　ID　　　ExtRPCIntRPC　State Role　　DsgBridgeDsgPort
--------------- ------- --------- --------- --- ---- ------------------ -------
 Ethernet0/0/3 128.003　　　　0　　　　0 FWD DSGN 32768.00030f0bf812 128.003

Ethernet0/0/4 128.004   0   0 FWD DSGN 32768.00030f0bf812 128.004

从结果中可以看出，交换机 B 是根交换机，交换机 A 的 1 端口是根端口。

任务四　多实例生成树

♂ 需求分析

相对于单实例生成树，多实例生成树允许多个具有相同拓扑的 VLAN 映射到一个生成树实例上，而这个生成树拓扑同其他生成树实例相互独立。这种多重生成树实例为映射到它的 VLAN 的数据流量提供了独立的发送路径，可实现不同实例间 VLAN 数据流量的负载分担。

♂ 方案设计

多实例生成树由于多个 VLAN 可以映射到一个单一的生成树实例上，IEEE 802.1s 委员会提出了多生成树域的概念，用来解决如何判断某个 VLAN 映射到哪个生成树实例的问题。在此任务环境中，将进一步理解多 VLAN 的生成树协议的原理和实际拓扑的生成。

所需设备如图 2-4-1 所示。
（1）DCRS-5650 交换机 2 台。
（2）PC 2 台。
（3）Console 线 1 或 2 条。
（4）直通网线 4～8 条。

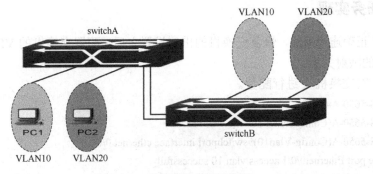

图 2-4-1　多实例生成树拓扑图

任务要求：如果多实例生成树成功，则通过"show spanning-tree mst"命令可观察到不同实例中 Trunk 链路的阻塞状况，使 VLAN10 只通过 23 口，VLAN20 只通过 24 口，用多实例生成树完成数据流量的负载均衡，见表 2-4-1。

表 2-4-1　多实例生成树配置表

设备 A	设备 B
switch A e0/0/23	switch B e0/0/23
switch A e0/0/24	switch B e0/0/24
PC1	switch A e0/0/1
PC2	switch B e0/0/9

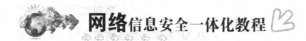

知识准备

1. 生成树协议

目的：为了防止冗余时产生环路。

原理：所有 VLAN 成员端口都加入一棵树，将备用链路的端口设为 BLOCK，主链路出现问题之后，BLOCK 的链路才成为 UP，端口的状态转换 BLOCK>LISTEN>LERARN>FORWARD>DISABLE 总共经历 50s，生成树协议工作时，正常情况下，交换机的端口要经过几个工作状态的转变。物理链路待接通时，将在 BLOCK 状态停留 20s，之后是 LISTEN 状态 15s，经过 15s 的 LERARN，最后成为 FORWARD 状态。

缺点：收敛速度慢，效率低。

解决收敛速度慢的补丁：POSTFACT/UPLINKFAST（检查直连链路）/BACKBONEFAST。

2. 多生成树协议

目的：解决 STP 与 RSTP 中效率低、占用资源的问题。

原理：部分 VLAN 为一棵树。

如果想在交换机上运行 MSTP，首先必须在全局下打开 MSTP 开关。在没有打开全局 MSTP 开关之前，打开端口的 MSTP 开关是不允许的。MSTP 定时器参数之间是有相关性的，错误配置可能会导致交换机不能正常工作。用户在修改 MSTP 参数时，应该清楚所产生的各个拓扑。除了全局的基于网桥的参数配置之外，其他的是基于各个实例的配置，在配置时一定要注意参数对应的实例是否正确。

任务实现

步骤 1：正确连接网线，恢复交换机的出厂设置之后，配置交换机的 VLAN 信息，配置端口到 VLAN 的映射关系。

步骤 2：对交换机 A 进行配置。

```
DCRS-5650-A#config
DCRS-5650-A(Config)#vlan 10
DCRS-5650-A(Config-Vlan10)#switchport interface ethernet 0/0/1-8
Set the port Ethernet0/0/1 access vlan 10 successfully
Set the port Ethernet0/0/2 access vlan 10 successfully
Set the port Ethernet0/0/3 access vlan 10 successfully
Set the port Ethernet0/0/4 access vlan 10 successfully
Set the port Ethernet0/0/5 access vlan 10 successfully
Set the port Ethernet0/0/6 access vlan 10 successfully
Set the port Ethernet0/0/7 access vlan 10 successfully
Set the port Ethernet0/0/8 access vlan 10 successfully
DCRS-5650-A(Config-Vlan10)#exit
DCRS-5650-A(Config)#vlan 20
DCRS-5650-A(Config-Vlan20)#switchport interface ethernet 0/0/9-16
Set the port Ethernet0/0/9 access vlan 20 successfully
Set the port Ethernet0/0/10 access vlan 20 successfully
```

Set the port Ethernet0/0/11 access vlan 20 successfully
Set the port Ethernet0/0/12 access vlan 20 successfully
Set the port Ethernet0/0/13 access vlan 20 successfully
Set the port Ethernet0/0/14 access vlan 20 successfully
Set the port Ethernet0/0/15 access vlan 20 successfully
Set the port Ethernet0/0/16 access vlan 20 successfully
DCRS-5650-A(Config-Vlan20)#exit
DCRS-5650-A(Config)#interface ethernet 0/0/23-24
DCRS-5650-A(Config-If-Port-Range)#switchport mode trunk
Set the port Ethernet0/0/23 mode TRUNK successfully
Set the port Ethernet0/0/24 mode TRUNK successfully
DCRS-5650-A(Config-If-Port-Range)#exit
DCRS-5650-A(Config)#

步骤3：对交换机B进行配置。

DCRS-5650-B#config
DCRS-5650-B(Config)#vlan 10
DCRS-5650-B(Config-Vlan10)#switchport interface ethernet 0/0/1-8
Set the port Ethernet0/0/1 access vlan 10 successfully
Set the port Ethernet0/0/2 access vlan 10 successfully
Set the port Ethernet0/0/3 access vlan 10 successfully
Set the port Ethernet0/0/4 access vlan 10 successfully
Set the port Ethernet0/0/5 access vlan 10 successfully
Set the port Ethernet0/0/6 access vlan 10 successfully
Set the port Ethernet0/0/7 access vlan 10 successfully
Set the port Ethernet0/0/8 access vlan 10 successfully
DCRS-5650-B(Config-Vlan10)#exit
DCRS-5650-B(Config)#vlan 20
DCRS-5650-B(Config-Vlan20)#switchport interface ethernet 0/0/9-16
Set the port Ethernet0/0/9 access vlan 20 successfully
Set the port Ethernet0/0/10 access vlan 20 successfully
Set the port Ethernet0/0/11 access vlan 20 successfully
Set the port Ethernet0/0/12 access vlan 20 successfully
Set the port Ethernet0/0/13 access vlan 20 successfully
Set the port Ethernet0/0/14 access vlan 20 successfully
Set the port Ethernet0/0/15 access vlan 20 successfully
Set the port Ethernet0/0/16 access vlan 20 successfully
DCRS-5650-B(Config-Vlan20)#exit
DCRS-5650-B(Config)#interface ethernet 0/0/23-24
DCRS-5650-B(Config-If-Port-Range)#switchport mode trunk
Set the port Ethernet0/0/23 mode TRUNK successfully
Set the port Ethernet0/0/24 mode TRUNK successfully
DCRS-5650-B(Config-If-Port-Range)#exit
DCRS-5650-B(Config)#

步骤4：配置多实例生成树，在交换机A上将VLAN 10映射到实例1上，将VLAN 20映射到实例 2 上。

```
DCRS-5650-A(Config)#spanning-tree mst configuration
DCRS-5650-A(Config-Mstp-Region)#name mstp
DCRS-5650-A(Config-Mstp-Region)#instance 1 vlan10
DCRS-5650-A(Config-Mstp-Region)#instance 2 vlan20
DCRS-5650-A(Config-Mstp-Region)#exit
DCRS-5650-A(Config)#spanning-tree
MSTP is starting now, please wait..........
MSTP is enabled successfully.
```

步骤 5：配置多实例生成树，在交换机 B 上将 VLAN 10 映射到实例 1 上，将 VLAN 20 映射到实例 2 上。

```
DCRS-5650-B(Config)#spanning-tree mst configuration
DCRS-5650-B(Config-Mstp-Region)#name mstp
DCRS-5650-B(Config-Mstp-Region)#instance 1 vlan10
DCRS-5650-B(Config-Mstp-Region)#instance 2 vlan20
DCRS-5650-B(Config-Mstp-Region)#exit
DCRS-5650-B(Config)#spanning-tree
MSTP is starting now, please wait..........
MSTP is enabled successfully.
```

步骤 6：查找根交换机。

```
switchA#show spanning-tree
                -- MSTPBridgeConfig Info --
Standard        :   IEEE 802.1s
Bridge MAC      :   00:03:0f:0b:f8:12
Bridge Times    :   Max Age 20, Hello Time 2, Forward Delay 15
Force Version:  3
########################## Instance 0 ##########################
SelfBridge Id       : 32768 -   00:03:0f:0b:f8:12
Root Id             : this switch
Ext.RootPathCost : 0
Region Root Id      : this switch
Int.RootPathCost : 0
Root Port ID        : 0
Current port list in Instance 0:
………………………
```

从结果中可以看出，交换机 A 是根交换机。

步骤 7：在根交换机上修改 Trunk 端口在不同实例中的优先级。

```
DCRS-5650-A(Config)#interface ethernet 0/0/23
DCRS-5650-A(Config-If-Ethernet0/0/23)#spanning-tree mst 1 port-priority 32
DCRS-5650-A(Config-If-Ethernet0/0/23)#exit
DCRS-5650-A(Config)#interface ethernet 0/0/24
DCRS-5650-A(Config-If-Ethernet0/0/24)#spanning-tree mst2 port-priority 32
DCRS-5650-A(Config-If-Ethernet0/0/24)#exit
```

项目二 VLAN 技术与生成树技术

DCRS-5650-A(Config)#

步骤 8：配置交换机 B 上各 VLAN 所属的 loopback 端口，保证各 VLAN 在线。

DCRS-5650-B(Config)#interface ethernet 0/0/1
DCRS-5650-B(Config-If-Ethernet0/0/1)#loopback
DCRS-5650-B(Config-If-Ethernet0/0/1)#exit
DCRS-5650-B(Config)#interface ethernet 0/0/9
DCRS-5650-B(Config-If-Ethernet0/0/9)#loopback
DCRS-5650-B(Config-If-Ethernet0/0/9)#exit

步骤 9：运行 "show spanning-tree mst" 命令，观察现象。

```
DCRS-5650-A#show spanning-tree mst
########################## Instance 0 ##########################
 vlans mapped       : 1-9;11-19;21-4094
 Self BridgeId      : 32768.00:03:0f:0b:f8:12
 Root Id            : this switch
 Root Times         : Max Age 20, Hello Time 2, Forward Delay 15 ,max hops 20
 PortName         ID       ExtRPC IntRPC  State Role   DsgBridge          DsgPort
 --------------   -------  ------ ------  ----- -----  ----------------   -------
  Ethernet0/0/1  128.001   0            0 FWD   DSGN   32768.00030f0bf812 128.001
  Ethernet0/0/9  128.009   0            0 FWD   DSGN   32768.00030f0bf812 128.009
  Ethernet0/0/23 128.023   0            0 FWD   DSGN   32768.00030f0bf812 128.023
  Ethernet0/0/24 128.024   0            0 FWD   DSGN   32768.00030f0bf812 128.024
########################## Instance 1 ##########################
 vlans mapped       : 10
 Self BridgeId      : 32768-00:03:0f:0b:f8:12
 Root Id            : this switch
 PortName         ID       IntRPC  State Role   DsgBridge          DsgPort
 --------------   -------  ------  ----- -----  ----------------   -------
  Ethernet0/0/1  128.001        0  FWD   DSGN   32768.00030f0bf812 128.001
  Ethernet0/0/23 032.023        0  FWD   DSGN   32768.00030f0bf812 032.023
  Ethernet0/0/24 128.024        0  FWD   DSGN   32768.00030f0bf812 128.024
########################## Instance 2 ##########################
 vlans mapped       : 20
 Self BridgeId      : 32768-00:03:0f:0b:f8:12
 Root Id            : this switch
 PortName         ID       IntRPC  State Role   DsgBridge          DsgPort
 --------------   -------  ------  ----- -----  ----------------   -------
  Ethernet0/0/9  128.009        0  FWD   DSGN   32768.00030f0bf812 128.009
  Ethernet0/0/23 128.023        0  FWD   DSGN   32768.00030f0bf812 128.023
  Ethernet0/0/24 032.024        0  FWD   DSGN   32768.00030f0bf812 032.024
########################## Instance 3 ##########################
DCRS-5650-B(config)#sh spanning-tree mst
########################## Instance 0 ##########################
 vlans mapped       : 1-9;11-19;21-4094
 Self BridgeId      : 32768.00:03:0f:0f:6e:ad
```

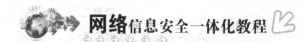

```
 Root Id              : 32768.00:03:0f:0b:f8:12
 Root Times           : Max Age 20, Hello Time 2, Forward Delay 15 ,max hops 19
 PortName        ID       ExtRPC IntRPC   State Role      DsgBridge DsgPort
 --------------- -------- -------- --- ---- ------------------  -------
  Ethernet0/0/1  128.001     0       200000 FWD DSGN 32768.00030f0f6ead 128.001
  Ethernet0/0/9  128.009     0       200000 FWD DSGN 32768.00030f0f6ead 128.009
  Ethernet0/0/22 128.022     0       200000 FWD DSGN 32768.00030f0f6ead 128.022
  Ethernet0/0/23 128.023     0            0 FWD ROOT 32768.00030f0bf812 128.023
  Ethernet0/0/24 128.024     0            0 BLK ALTR 32768.00030f0bf812 128.024
 ########################## Instance 1 ##########################
 vlans mapped       : 10
 Self BridgeId      : 32768-00:03:0f:0f:6e:ad
 Root Id            : 32768.00:03:0f:0b:f8:12
 PortName        ID       IntRPC     State Role      DsgBridge DsgPort
 --------------- -------- --------- --- ---- ------------------  -------
  Ethernet0/0/1  128.001   200000 FWD DSGN 32768.00030f0f6ead 128.001
  Ethernet0/0/23 128.023        0 FWD ROOT 32768.00030f0bf812 032.023
  Ethernet0/0/24 128.024        0 BLK ALTR 32768.00030f0bf812 128.024
 ########################## Instance 2 ##########################
 vlans mapped       : 20
 Self BridgeId      : 32768-00:03:0f:0f:6e:ad
 Root Id            : 32768.00:03:0f:0b:f8:12
 PortName        ID       IntRPC     State Role      DsgBridge DsgPort
 --------------- -------- --------- --- ---- ------------------  -------
  Ethernet0/0/9  128.009   200000 FWD DSGN 32768.00030f0f6ead 128.009
  Ethernet0/0/23 128.023        0 BLK ALTR 32768.00030f0bf812 128.023
  Ethernet0/0/24 128.024        0 FWD ROOT 32768.00030f0bf812 032.024
```

任务五　改变生成树状态

♂ 需求分析

使用生成树默认参数一般很难使管理员得到希望的网络拓扑。在规划一个网络时，管理员有自己预期的拓扑，如希望业务部门的数据流从某一个骨干链路传递，而客户服务部门的数据从另一个骨干链路传递，这两个链路可以同时提供互为备份的关系，当其中业务部门的骨干链路断开时，其数据也可以从客户服务部门使用的链路通过，这样虽然拥挤但是可以保证数据的实时传递效果。

♂ 方案设计

使用生成树协议可使这样的愿望得以实现。

本任务使用 DCS-3926S 系列交换机作为演示设备，其软件版本为 DCS-3926S_6.1.12.0，实际使用中由于软件版本不同,功能和配置方法有可能存在差异,请关注相应版本的使用说明。

所需设备如图 2-5-1 所示。

(1) DCRS-5650 交换机 2 台。
(2) PC 2 台。
(3) Console 线 1 或 2 条。
(4) 直通网线 4~8 条。

图 2-5-1 改变生成树状态

任务要求:多实例生成树成功后,通过"show spanning-tree mst"命令查看不同实例中 Trunk 链路的阻塞状况,通过配置可以改变端口的阻塞情况,如表 2-5-1 所示。

表 2-5-1 多实例生成树配置表

设备 A	设备 B
交换机 A e0/0/23	交换机 B e0/0/23
交换机 A e0/0/24	交换机 B e0/0/24
PC1	交换机 A e0/0/1
PC2	交换机 B e0/0/9

步骤 1:基本配置同本项目任务四,此处略。

步骤 2:通过改变交换机 A 的端口优先级实现生成树拓扑形态的改变。这里选择将交换机 A 的 1 端口优先级升高,改变拓扑。

```
switchA#config
switchA(Config)#interface ethernet 0/0/2
switchA(Config-Ethernet0/0/2)#spanning-tree mst 0 port-priority 112
//改变接口优先级
switchA(Config-Ethernet0/0/2)#
```

步骤 3:查看 switchA 的生成树配置。

```
switchA#show spanning-tree
                -- MSTPBridgeConfig Info --
Standard        :   IEEE 802.1s
Bridge MAC      :   00:03:0f:00:5d:50
Bridge Times :    Max Age 20, Hello Time 2, Forward Delay 15
Force Version:    3
########################## Instance 0 ##########################
SelfBridge Id    : 32768 -    00:03:0f:00:5d:50
Root Id             : this switch
Ext.RootPathCost : 0
Region Root Id    : this switch
Int.RootPathCost : 0
Root Port ID       : 0
Current port list in Instance 0:
```

```
    Ethernet0/0/1 Ethernet0/0/2 (Total 2)
    PortName          ID       ExtRPC IntRPC  State Role   DsgBridge DsgPort
    ---------------   ------   ------ ------  ----- ----   -----------------  -------
      Ethernet0/0/1  144.001     0           0 FWD DSGN  32768.00030f005d50  144.001
      Ethernet0/0/2  128.002     0           0 FWD DSGN  32768.00030f005d50  128.002
    switchA#
```

步骤 4：查看 switchB 的生成树配置。

```
    switchB#show spanning-tree
                   -- MSTPBridgeConfig Info --
    Standard       :  IEEE 802.1s
    Bridge MAC     :  00:03:0f:01:ec:0a
    Bridge Times   :  Max Age 20, Hello Time 2, Forward Delay 15
    Force Version  :  3
    ########################### Instance 0 ###########################
    SelfBridge Id    : 32768 -  00:03:0f:01:ec:0a
    Root Id          : 32768.00:03:0f:00:5d:50
    Ext.RootPathCost : 200000
    Region Root Id   : this switch
    Int.RootPathCost : 0
    Root Port ID     : 128.3
    Current port list in Instance 0:
    Ethernet0/0/3 Ethernet0/0/4 (Total 2)
    PortName          ID       ExtRPC IntRPC  State Role   DsgBridge DsgPort
    ---------------   ------   ------ ------  ----- ----   -----------------  -------
      Ethernet0/0/3  128.003     0           0 FWD ROOT  32768.00030f005d50  128.002
      Ethernet0/0/4  128.004     0           0 BLK ALTR  32768.00030f005d50  144.001
    switchB#
```

此时，交换机 B 的 4 端口已经更新为阻塞状态，拓扑形态已经改变，如图 2-5-2 所示。

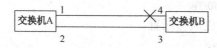

图 2-5-2　交换机 B 端口 4 的阻塞状态

步骤 5：通过改变交换机 B 的优先级实现生成树拓扑形态的改变。

```
    switchB#config
    switchB(Config)#spanning-tree mst 0 priority 28672  //改变交换机的优先级
    switchB(Config)#exit
    switchB#show span
                   -- MSTPBridgeConfig Info --
    Standard       :  IEEE 802.1s
    Bridge MAC     :  00:03:0f:01:ec:0a
    Bridge Times   :  Max Age 20, Hello Time 2, Forward Delay 15
    Force Version  :  3
    ########################### Instance 0 ###########################
    SelfBridge Id    : 28672 -  00:03:0f:01:ec:0a
```

```
Root Id              : this switch
Ext.RootPathCost : 0
Region Root Id       : this switch
Int.RootPathCost : 0
Root Port ID         : 0
Current port list in Instance 0:
Ethernet0/0/3 Ethernet0/0/4 (Total 2)
PortName        ID      ExtRPCIntRPC   State Role    DsgBridgeDsgPort
-------------- ------- ---------- --- ---- ------------------- --------
  Ethernet0/0/3 128.003        0          0 FWD DSGN 28672.00030f01ec0a 128.003
  Ethernet0/0/4 128.004        0          0 FWD DSGN 28672.00030f01ec0a 128.004
switchB#
switchA#show spanning-tree
                       -- MSTPBridgeConfig Info --
Standard       :  IEEE 802.1s
Bridge MAC     :  00:03:0f:00:5d:50
Bridge Times : Max Age 20, Hello Time 2, Forward Delay 15
Force Version :   3
######################### Instance 0 #########################
SelfBridge Id : 32768 -  00:03:0f:00:5d:50
Root Id              : 28672.00:03:0f:01:ec:0a
Ext.RootPathCost : 200000
Region Root Id       : this switch
Int.RootPathCost : 0
Root Port ID         : 128.2
Current port list in Instance 0:
Ethernet0/0/1 Ethernet0/0/2 (Total 2)
PortName        ID      ExtRPCIntRPC   State Role    DsgBridgeDsgPort
-------------- ------- ---------- --- ---- ------------------- --------
  Ethernet0/0/1 128.001        0          0 BLK ALTR 28672.00030f01ec0a 128.004
  Ethernet0/0/2 128.002        0          0 FWD ROOT 28672.00030f01ec0a 128.003
switchA#
```

此时,交换机 A 的 1 端口已经更新为阻塞状态,拓扑形态已经改变,如图 2-5-3 所示。

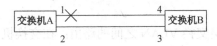

图 2-5-3　交换机 A 端口 1 的阻塞状态

小贴士

(1)命令中使用了 mst 关键字,它是多实例生成树英文名词的缩写,其后跟的"0"代表这里对第 0 个实例的参数进行更改。交换机默认时,其生成树协议数据单元都从 VLAN 1 发送,而且将所有 VLAN 都对应到生成树的实例 0 中,因此,对交换机没有使用多实例设置时,可以使用实例 0 更改其参数。

(2)在端口优先级的修改中,系统支持以 16 为基数进行增减;在设备优先级的修改中,系统支持以 4096 为基数进行增减,因此,在以上的配置中均以默认值增加 16 和默认值减少 4096

为例进行配置。

认证考核

一、选择题

1. VLAN 在现代组网技术中占有重要地位。在由多个 VLAN 组成的一个局域网中，以下说法不正确的是（ ）。
   A．当站点从一个 VLAN 转移到另一个 VLAN 时，一般不需要改变物理连接
   B．VLAN 中的一个站点可以和另一个 VLAN 中的站点直接通信
   C．当站点在一个 VLAN 中广播时，其他 VLAN 中的站点不能收到数据报
   D．VLAN 可以通过 MAC 地址、交换机端口等进行定义

2. 在默认配置的情况下，交换机的所有端口（ ）。
   A．处于直通状态   B．属于同一 VLAN
   C．属于不同 VLAN   D．地址都相同

3. 连接在不同交换机上的，属于同一 VLAN 的数据帧必须通过（ ）传输。
   A．服务器   B．路由器   C．Backbone 链路   D．Trunk 链路

4. 下面关于 VLAN 的描述中，不正确的是（ ）。
   A．VLAN 把交换机划分成多个逻辑上独立的交换机
   B．主干链路（Trunk）可以提供多个 VLAN 之间通信的公共通道
   C．由于包含了多个交换机，所以 VLAN 扩大了冲突域
   D．一个 VLAN 可以跨越多个交换机

5. 当数据在两个 VLAN 之间传输时需要使用（ ）。
   A．二层交换机   B．网桥   C．路由器   D．中继器

二、操作题

使用三层交换机实现二层交换 VLAN 之间的路由：在交换机 A 和交换机 B 上分别划分两个基于端口的 VLAN——VLAN10、VLAN20，见表 2-5-2。

表 2-5-2 交换机 A、B 的 VLAN 划分、Trunk 口设置

VLAN	端口成员
10	2～4
20	5～8
Trunk 口	1

在交换机 C 上划分两个基于端口的 VLAN——VLAN10、VLAN20。把端口 1 和端口 2 都设置成 Trunk 口，见表 2-5-3。

项目二
VLAN 技术与生成树技术

表 2-5-3 交换机 C 的 VLAN 划分、Trunk 口设置

VLAN	IP 地址	子网掩码
10	10.1.10.1	255.255.255.0
20	10.1.20.1	255.255.255.0
Trunk 口		1 和 2

交换机 A 的端口 1 连接交换机 C 的端口 1，交换机 B 的端口 1 连接交换机 C 的端口 2。PC 的网络设置见表 2-5-4，现要求 PC1 可以 ping 通 PC2。

表 2-5-4 PC1 和 PC2 网络参数设置

设备	端口	IP 地址	网关	子网掩码
PC1	switchA 端口 2	10.1.10.11	10.1.10.1	255.255.255.0
PC2	switchB 端口 8	10.1.20.22	10.1.20.1	255.255.255.0

项目三 路由协议

教学背景

在实际应用中路由器通常连接着许多不同的网络，要实现多个不同网络间的通信，就需要在路由器上配置路由协议。路由器提供的路由协议包括静态路由协议、RIP 动态路由协议、OSPF 动态路由协议等。

任务一　实现静态路由

需求分析

某公司刚成立，规模很小，只有两台路由器。该公司的网络管理员经过考虑，决定在公司的路由器与运营商路由器之间使用静态路由，实现网络的互连。

方案设计

由于该网络规模较小且不经常变动，因此使用静态路由比较合适。两台路由器之间通过以太网口相连，每台路由器连接一台计算机。在小规模环境里，静态路由是最佳的选择。静态路由开销小，但不灵活，适用于相对稳定的网络

所需设备如图 3-1-1 所示。

（1）DCR-2626 路由器 3 台。
（2）CR-V35FC 1 条。
（3）CR-V35MT 1 条。

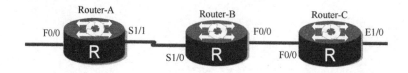

图 3-1-1　实现静态路由拓扑图

各路由器接口配置要求见表 3-1-1。

表 3-1-1 设备配置信息表

Router-A		Router-B		Router-C	
接口	IP 地址	接口	IP 地址	接口	IP 地址
S1/1(DCE)	192.168.1.1	S/1/0(DTE)	192.168.1.2	F0/0	192.168.2.2
F0/0	192.168.0.1	F0/0	192.168.2.1	E1/0	192.168.3.1

知识准备

1. 路由表的产生方式

路由器在转发数据时，首先要在路由表中查找相应的路由。路由表的产生方式有以下 3 种。
（1）直连网络：路由器自动添加和自己直接连接的网络路由。
（2）静态路由：由网络管理员手工配置的路由信息。当网络的拓扑结构或链路发生变化时，需要网络管理员手动修改路由表中的相关路由信息。
（3）动态路由：由路由协议动态产生的路由。

2. 静态路由的优缺点

静态路由的优点：使用静态路由的一个好处是网络保密性高。动态路由因为需要路由器之间频繁地交换各自的路由表，而对路由表的分析可以揭示网络的拓扑结构和网络地址等信息，因此，出于网络安全方面的考虑可以采用静态路由。

静态路由的缺点：大型和复杂的网络环境通常不宜采用静态路由。一方面，网络管理员难以全面地了解整个网络的拓扑结构；另一方面，当网络的拓扑结构和链路状态发生变化时，路由器中的静态路由信息需要大范围地调整，这一工作的难度和复杂程度非常高。

在小型的网络中，使用静态路由是较好的选择，管理员想控制数据转发路径时，也会使用静态路由。

静态路由的配置命令如下。

ip route 目标网络的 IP 地址 子网掩码 下一跳 IP 地址/本地接口

任务实现

步骤 1：按照表 3-1-1 配置所有接口的 IP 地址，保证所有接口全部是 up 状态，测试连通性。

步骤 2：查看 Router-A 的路由表。

```
Router-A#show ip route
Codes: C - connected, S - static, R - RIP, B - BGP, BC - BGP connected
       D - DEIGRP, DEX - external DEIGRP, O - OSPF, OIA - OSPF inter area
       ON1 - OSPF NSSA external type 1, ON2 - OSPF NSSA external type 2
       OE1 - OSPF external type 1, OE2 - OSPF external type 2
       DHCP - DHCP type
VRF ID: 0
C      192.168.0.0/24       is directly connected, FastEthernet0/0//直连的路由
C      192.168.1.0/24       is directly connected, Serial1/1    //直连的路由
```

步骤 3：查看 Router-B 的路由表。

```
Router-B#show ip route
Codes: C - connected, S - static, R - RIP, B - BGP, BC - BGP connected
       D - DEIGRP, DEX - external DEIGRP, O - OSPF, OIA - OSPF inter area
       ON1 - OSPF NSSA external type 1, ON2 - OSPF NSSA external type 2
       OE1 - OSPF external type 1, OE2 - OSPF external type 2
       DHCP - DHCP type
VRF ID: 0
   C     192.168.1.0/24          is directly connected, Serial1/0
   C     192.168.2.0/24          is directly connected, FastEthernet0/0
```

步骤 4：查看 Router-C 的路由表。

```
Router-B#show ip route
Codes: C - connected, S - static, R - RIP, B - BGP, BC - BGP connected
       D - DEIGRP, DEX - external DEIGRP, O - OSPF, OIA - OSPF inter area
       ON1 - OSPF NSSA external type 1, ON2 - OSPF NSSA external type 2
       OE1 - OSPF external type 1, OE2 - OSPF external type 2
       DHCP - DHCP type
VRF ID: 0
   C     192.168.1.0/24          is directly connected, Serial1/0
   C     192.168.2.0/24          is directly connected, FastEthernet0/0
```

步骤 5：在 Router-A 上 ping 路由器 C。

```
Router-A#ping 192.168.2.2
PING 192.168.2.2 (192.168.2.2): 56 data bytes
.....
--- 192.168.2.2 ping statistics ---
5 packets transmitted, 0 packets received, 100% packet loss          //不通
```

步骤 6：在 Router-A 上配置静态路由。

```
Router-A#config
Router-A_config#ip route 192.168.2.0 255.255.255.0 192.168.1.2
//配置目标网段和下一跳 IP 地址
Router-A_config#ip route 192.168.3.0 255.255.255.0 192.168.1.2
```

步骤 7：查看 Router-A 的路由表。

```
Router-A#show ip route
Codes: C - connected, S - static, R - RIP, B - BGP, BC - BGP connected
       D - DEIGRP, DEX - external DEIGRP, O - OSPF, OIA - OSPF inter area
       ON1 - OSPF NSSA external type 1, ON2 - OSPF NSSA external type 2
       OE1 - OSPF external type 1, OE2 - OSPF external type 2
       DHCP - DHCP type
VRF ID: 0
   C     192.168.0.0/24          is directly connected, FastEthernet0/0
   C     192.168.1.0/24          is directly connected, Serial1/1
   S     192.168.2.0/24          [1,0] via 192.168.1.2//注意，静态路由的管理距离是 1
   S     192.168.3.0/24          [1,0] via 192.168.1.2
```

步骤 8：配置 Router-B 的静态路由并查看路由表。

```
Router-B#config
```

```
Router-B_config#ip route 192.168.0.0 255.255.255.0 192.168.1.1
Router-B_config#ip route 192.168.3.0 255.255.255.0 192.168.2.2
Router-B_config#^Z
Router-B#show ip route
Codes: C - connected, S - static, R - RIP, B - BGP, BC - BGP connected
       D - DEIGRP, DEX - external DEIGRP, O - OSPF, OIA - OSPF inter area
       ON1 - OSPF NSSA external type 1, ON2 - OSPF NSSA external type 2
       OE1 - OSPF external type 1, OE2 - OSPF external type 2
       DHCP - DHCP type
VRF ID: 0
S       192.168.0.0/24       [1,0] via 192.168.1.1
C       192.168.1.0/24       is directly connected, Serial1/0
C       192.168.2.0/24       is directly connected, FastEthernet0/0
S       192.168.3.0/24       [1,0] via 192.168.2.2
```

步骤 9：配置 Router-C 的静态路由并查看路由表。

```
Router-C#config
Router-C_config#ip route 192.168.0.0 255.255.0.0 192.168.2.1  //采用了超网的方法
Router-C_config#^Z
Router-C#show ip route
Codes: C - connected, S - static, R - RIP, B - BGP
       D - DEIGRP, DEX - external DEIGRP, O - OSPF, OIA - OSPF inter area
       ON1 - OSPF NSSA external type 1, ON2 - OSPF NSSA external type 2
       OE1 - OSPF external type 1, OE2 - OSPF external type 2
S       192.168.0.0/16       [1,0] via 192.168.2.1        //注意，子网掩码是 16 位
C       192.168.2.0/24       is directly connected,   FastEthernet0/0
C       192.168.3.0/24       is directly connected,   Ethernet1/0
```

步骤 10：测试。

```
Router-C#ping 192.168.0.1
PING 192.168.0.1 (192.168.0.1): 56 data bytes
!!!!!                                                              //成功
--- 192.168.0.1 ping statistics ---
5 packets transmitted, 5 packets received, 0% packet loss
round-trip min/avg/max = 30/32/40 ms
```

任务二　实现 RIP 基本配置

需求分析

随着公司规模的不断扩大，路由器的数量已经达到了 5 台。该公司的网络管理员发现原有的静态路由已经不适合现在的公司网络了，因此，他决定在公司的路由器之间使用动态的 RIP 协议，实现网络的互连。

方案设计

在路由器较多的环境里，手工配置静态路由给网络管理员带来了很大的工作负担。此外，在不太稳定的网络环境里，手工修改路由表也不现实。

所需设备如图 3-2-1 所示。

（1）DCR-2626 路由器 3 台。
（2）CR-V35FC 1 条。
（3）CR-V35MT 1 条。

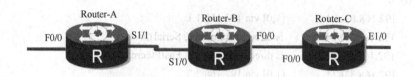

图 3-2-1 实现 RIP 基本配置

各路由器接口配置要求见表 3-2-1。

表 3-2-1 各路由器接口配置表

Router-A		Router-B		Router-C	
接口	IP 地址	接口	IP 地址	接口	IP 地址
S1/1(DCE)	192.168.1.1	S/1/0(DTE)	192.168.1.2	F0/0	192.168.2.2
F0/0	192.168.0.1	F0/0	192.168.2.1	E1/0	192.168.3.1

知识准备

RIP（Routing Information Protocol，路由信息协议）采用了距离矢量算法。在默认情况下，RIP 使用一种非常简单的度量制度：距离就是通往目的站点所需经过的链路数，取值为 1～15，数值 16 表示无穷大。RIP 进程使用 UDP 的 520 端口来发送和接收 RIP 分组。RIP 分组每隔 30s 以广播的形式发送一次，为了防止出现"广播风暴"，其后续的分组将做随机延时后再发送。在 RIP 中，如果一个路由在 180s 内未被刷新，则相应的距离就被设定成无穷大，并从路由表中删除该表项。RIP 分组有两种：请求分组和响应分组。

RIPv1 被提出得较早，其中有许多缺陷。为了改善 RIPv1 的不足，在 RFC 1388 中提出了改进的 RIPv2，并在 RFC 1723 和 RFC 2453 中进行了修订。RIPv2 定义了一套有效的改进方案，新的 RIPv2 支持子网路由选择，支持 CIDR，支持组播，并提供了验证机制。

RIP 有以下主要特性。

（1）RIP 属于典型的距离矢量路由选择协议。

（2）RIP 消息通过广播地址 255.255.255.255 进行发送，RIPv2 使用组播地址 224.0.0.9 发送消息，两者都使用 UDP 的 520 端口。

（3）RIP 以到达目的网络的最小跳数作为路由选择度量标准，而不是在链路的带宽和延迟的基础上进行选择。

（4）RIP 是为小型网络设计的。它的跳数计数限制为 15 跳，16 跳为不可到达。

（5）RIPv1 是一种有类路由协议，不支持不连续子网设计。RIPv2 支持 CIDR 及可变长子网掩码，因此支持不连续子网设计。

（6）RIP 周期性进行完全路由更新，将路由表广播给邻居路由器，广播周期默认为 30s。

（7）RIP 协议的管理距离为 120。

任务实现

步骤 1：参照前面的任务，按照表 3-2-1 配置所有接口的 IP 地址，保证所有接口全部是 up 状态，测试连通性。

步骤 2：在 Router-A 上 ping RouterC。

```
Router-A#ping 192.168.2.2
PING 192.168.2.2 (192.168.2.2): 56 data bytes
......
--- 192.168.2.2 ping statistics ---
5 packets transmitted, 0 packets received, 100% packet loss      //不通
```

步骤 3：在 Router-A 上配置 RIP 并查看路由表。

```
Router-A_config#router rip                              //启动 RIP
Router-A_config_rip#network 192.168.0.0                 //宣告网段
Router-A_config_rip#network 192.168.1.0
Router-A_config_rip#^Z
Router-A#sh ip route
Codes: C - connected, S - static, R - RIP, B - BGP, BC - BGP connected
       D - DEIGRP, DEX - external DEIGRP, O - OSPF, OIA - OSPF inter area
       ON1 - OSPF NSSA external type 1, ON2 - OSPF NSSA external type 2
       OE1 - OSPF external type 1, OE2 - OSPF external type 2
       DHCP - DHCP type
VRF ID: 0
C       192.168.0.0/24      is directly connected, FastEthernet0/0
C       192.168.1.0/24      is directly connected, Serial1/1
```

注意，此时并没有出现 RIP 学习到的路由。

步骤 4：在 Router-B 上配置 RIP 协议并查看路由表。

```
Router-B_config#router rip
Router-B_config_rip#network 192.168.1.0
Router-B_config_rip#network 192.168.2.0
Router-B_config_rip#^Z
Router-B#2004-1-1 00:15:58 Configured from console 0 by DEFAULT
Router-B#show ip route
Codes: C - connected, S - static, R - RIP, B - BGP, BC - BGP connected
       D - DEIGRP, DEX - external DEIGRP, O - OSPF, OIA - OSPF inter area
       ON1 - OSPF NSSA external type 1, ON2 - OSPF NSSA external type 2
       OE1 - OSPF external type 1, OE2 - OSPF external type 2
       DHCP - DHCP type
VRF ID: 0
```

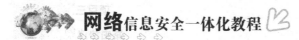

```
R       192.168.0.0/16      [120,1] via 192.168.1.1(on Serial1/0)    //从 A 学习到的路由
C       192.168.1.0/24      is directly connected, Serial1/0
C       192.168.2.0/24      is directly connected, FastEthernet0/0
```

步骤 5：在 Router-C 上配置 RIP 协议并查看路由表。

```
Router-C_config#router rip
Router-C_config_rip#network 192.168.2.0
Router-C_config_rip#network 192.168.3.0
Router-C_config_rip#^Z
Router-C#show ip route
Codes: C - connected, S - static, R - RIP, B - BGP
       D - DEIGRP, DEX - external DEIGRP, O - OSPF, OIA - OSPF inter area
       ON1 - OSPF NSSA external type 1, ON2 - OSPF NSSA external type 2
       OE1 - OSPF external type 1, OE2 - OSPF external type 2
R       192.168.0.0/16      [120,2] via 192.168.2.1(on    FastEthernet0/0)
R       192.168.1.0/24      [120,1] via 192.168.2.1(on    FastEthernet0/0)
C       192.168.2.0/24      is directly connected,    FastEthernet0/0
C       192.168.3.0/24      is directly connected,    Ethernet1/0
```

步骤 6：再次查看 Router-B 和 Router-A 的路由表。

```
Router-B#show ip route
Codes: C - connected, S - static, R - RIP, B - BGP, BC - BGP connected
       D - DEIGRP, DEX - external DEIGRP, O - OSPF, OIA - OSPF inter area
       ON1 - OSPF NSSA external type 1, ON2 - OSPF NSSA external type 2
       OE1 - OSPF external type 1, OE2 - OSPF external type 2
       DHCP - DHCP type
VRF ID: 0
R       192.168.0.0/16      [120,1] via 192.168.1.1(on Serial1/0)
C       192.168.1.0/24      is directly connected, Serial1/0
C       192.168.2.0/24      is directly connected, FastEthernet0/0
R       192.168.3.0/24      [120,1] via 192.168.2.2(on FastEthernet0/0)
Router-A#show ip route
Codes: C - connected, S - static, R - RIP, B - BGP, BC - BGP connected
       D - DEIGRP, DEX - external DEIGRP, O - OSPF, OIA - OSPF inter area
       ON1 - OSPF NSSA external type 1, ON2 - OSPF NSSA external type 2
       OE1 - OSPF external type 1, OE2 - OSPF external type 2
       DHCP - DHCP type
VRF ID: 0
C       192.168.0.0/24      is directly connected, FastEthernet0/0
C       192.168.1.0/24      is directly connected, Serial1/1
R       192.168.2.0/24      [120,1] via 192.168.1.2(on Serial1/1)
R       192.168.3.0/24      [120,2] via 192.168.1.2(on Serial1/1)
```

注意，此时所有网段都学习到了路由。

步骤 7：查看 RIP 状态。

```
Router-A#show ip rip                                //显示 RIP 状态
RIP protocol:   Enabled
  Global version: default( Decided on the interface version control )
```

Update: 30,　Expire: 180,　Holddown: 120
　　Input-queue: 50
　　Validate-update-source enable
　　No neighbor

步骤 8：显示协议细节。

　　Router-A#sh ip rip protocol　　　　　　　　　　//显示协议细节
　　RIP is Active
　　　Sending updates every 30 seconds, next due in 30 seconds　　//注意定时器的值
　　　Invalid after 180 seconds, holddown 120
　　update filter list for all interfaces is:
　　update offset list for all interfaces is:
　　Redistributing:
　　Default version control: send version 1, receive version 1 2
　　　Interface　　　　　Send　　　　　　Recv
　　　FastEthernet0/0　　1　　　　　　　1 2
　　　Serial1/1　　　　　1　　　　　　　1 2
　　Automatic network summarization is in effect
　　Routing for Networks:
　　　192.168.1.0/24
　　　192.168.0.0/16
　　Distance: 120 (default is 120)　　　　　　　　　//注意默认的管理距离
　　Maximum route count: 1024,　　Route count:6

步骤 9：显示 RIP 数据库。

　　Router-A#show ip rip database　　　　　　　　　//显示 RIP 数据库
　　192.168.0.0/24　directly connected　　FastEthernet0/0
　　192.168.0.0/24　auto-summary
　　192.168.1.0/24　directly connected　　Serial1/1
　　192.168.1.0/24　auto-summary
　　192.168.2.0/24　[120,1]　via 192.168.1.2 (on Serial1/1) 00:00:13
　　//收到 RIP 广播的时间
　　192.168.3.0/24　[120,2]　via 192.168.1.2 (on Serial1/1)　00:00:13

步骤 10：显示 RIP 路由。

　　Router-A#sh ip route rip　　　　　　　　　　　//仅显示 RIP 学习到的路由
　　R　　192.168.2.0/24　　　[120,1] via 192.168.1.2(on Serial1/1)
　　R　　192.168.3.0/24　　　[120,2] via 192.168.1.2(on Serial1/1)

任务三　实现 RIPv1 与 RIPv2 的兼容

需求分析

　　某公司刚成立，规模很小，只有两台路由器。由于业务发展，公司又新买了一台路由器，但前面的两台路由器已经配置成 RIPv1，现由于公司需要，使用了不规则的子网，网络管理员

经过考虑，决定几台路由器同时使用两个版本，既以 RIPv1 和 RIPv2 兼容的形式来实现网络的互连。

方案设计

网络协议的设计总是提供向后和向前的兼容性的，在 RIPv1 和 RIPv2 共存的环境中，通常也可以使用配置的方式进行兼容，而不必统一调整为 RIPv1 或版本 RIPv2。

所需设备如图 3-3-1 所示。

（1）DCR-2626 路由器 2 台。
（2）PC 2 台。
（3）网线 3 条。

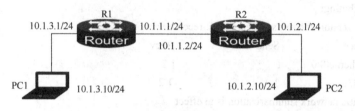

图 3-3-1　实现 RIPv1 与 RIPv2 的兼容

任务要求：配置基础环境，配置 R1 使用 RIPv1，R2 使用 RIPv2，分别使用 network 命令进行网段宣告，在 R2 中使用命令兼容 R1 的 RIPv1，开启 Debug 功能查看 R1 和 R2 收发的 RIP 报文。

知识准备

（1）RIPv1 是有类路由协议，RIPv2 是无类路由协议。
（2）RIPv1 不能支持 VLSM，RIPv2 可以支持 VLSM。
（3）RIPv1 没有认证的功能，RIPv2 可以支持认证，并且有明文和 MD5 两种认证。
（4）RIPv1 没有手工汇总的功能，RIPv2 可以在关闭自动汇总的前提下，进行手工汇总。
（5）RIPv1 是广播更新，RIPv2 是组播更新。

RIPv2 的特性如下。

① RIPv2 是一种无类路由协议。
② RIPv2 协议报文中携带了掩码信息，支持可变长子网掩码和 CIDR。
③ RIPv2 支持以组播方式发送路由更新报文，组播地址为 224.0.0.9，减少网络与系统资源消耗。
④ RIPv2 支持对协议报文进行验证，并提供明文验证和 MD5 验证两种方式，增强了安全性。

任务实现

步骤 1：配置路由器 R1 的基础网络环境。

```
Router_config#hostname R1
R1_config#interface fastEthernet 0/0
R1_config_f0/0#ip address 10.1.1.1 255.255.255.0
R1_config_f0/0#exit
```

```
R1_config#interface fastEthernet 0/1
R1_config_f0/1#ip address 10.1.3.1 255.255.255.0
R1_config_f0/1#exit
R1_config#
```

步骤2：配置路由器 R2 的基础网络环境。

```
Router_config#hostname R2
R2_config#interface fastEthernet 0/0
R2_config_f0/0#ip address 10.1.1.2 255.255.255.0
R2_config_f0/0#exit
R2_config#interface fastEthernet 0/3
R2_config_f0/3#ip address 10.1.2.1 255.255.255.0
R2_config_f0/3#exit
R2_config#
```

步骤3：在路由器 R1 和 R2 上配置 RIP 协议。

```
R1_config#router rip
R1_config_rip#network 10.1.1.0
R1_config_rip#network 10.1.3.0
R1_config_rip#version 1
R1_config_rip#exit
R2_config#router rip
R2_config_rip#network 10.1.1.0 255.255.255.0
R2_config_rip#network 10.1.2.0 255.255.255.0
R2_config_rip#version 2
R2_config_rip#
```

步骤4：查看 R1 的路由表。

```
R1_config#show ip route
Codes: C - connected, S - static, R - RIP, B - BGP, BC - BGP connected
       D - DEIGRP, DEX - external DEIGRP, O - OSPF, OIA - OSPF inter area
       ON1 - OSPF NSSA external type 1, ON2 - OSPF NSSA external type 2
       OE1 - OSPF external type 1, OE2 - OSPF external type 2
       DHCP - DHCP type

VRF ID: 0
C     10.1.1.0/24         is directly connected, FastEthernet0/0
R     10.1.2.0/24         [120,1] via 10.1.1.2(on FastEthernet0/0)
C     10.1.3.0/24         is directly connected, FastEthernet0/1
R1_config#
```

步骤5：查看 R2 的路由表。

```
R2#sh ip route
Codes: C - connected, S - static, R - RIP, B - BGP, BC - BGP connected
       D - DEIGRP, DEX - external DEIGRP, O - OSPF, OIA - OSPF inter area
       ON1 - OSPF NSSA external type 1, ON2 - OSPF NSSA external type 2
       OE1 - OSPF external type 1, OE2 - OSPF external type 2
       DHCP - DHCP type

VRF ID: 0
```

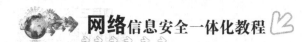

```
C        10.1.1.0/24              is directly connected, FastEthernet0/0
C        10.1.2.0/24              is directly connected, FastEthernet0/3
R        10.1.3.0/24              [120,1] via 10.1.1.1(on FastEthernet0/0)
R2#
```

步骤 6：这里可以观察到，即使 R1 使用了 RIPv1，而 R2 使用了 RIPv2，它们依然可以建立起路由表，互通性测试也是没有问题的，但详细查看 RIP 的数据库就有问题了，查看数据库后如下。

```
R2#sh ip rip data
  10.0.0.0/8           auto-summary
  10.1.1.0/24          directly connected    FastEthernet0/0
  10.1.2.0/24          directly connected    FastEthernet0/3
  10.1.3.0/24          [120,1]    via 10.1.1.1 (on FastEthernet0/0)    00:02:39
  ……//省略一段时间
R2#sh ip rip data
  10.0.0.0/8           auto-summary
  10.1.1.0/24          directly connected    FastEthernet0/0
  10.1.2.0/24          directly connected    FastEthernet0/3
  10.1.3.0/24          [120,16]   via 10.1.1.1 holddown (on FastEthernet0/0)   00:00:14
```

以上选取了 R2 的数据库查看，发现其对于远端网络的学习已经终止了，3min 后进入了 holddown 时间，再经过 2min 即从数据库中清除了这条路由。

而从 holddown 时间开始，从终端发起的测试连通就已经无法连通了，如下所示。

```
Reply from 10.1.3.10: bytes=32 time<1ms TTL=126
Reply from 10.1.3.10: bytes=32 time<1ms TTL=126
Reply from 10.1.2.1: Destination host unreachable.
Reply from 10.1.2.1: Destination host unreachable.
```

以上的实验表明 RIPv1 和 RIPv2 之间并不是自动兼容的。开启 R1 和 R2 的 Debug 功能可以得到如下信息。

```
R2#2002-1-1 01:56:26 RIP: send to 224.0.0.9 via FastEthernet0/0
2002-1-1 01:56:26       vers 2, CMD_RESPONSE, length 24
2002-1-1 01:56:26       10.1.2.0/24 via 0.0.0.0 metric 1
2002-1-1 01:56:26 RIP: send to 224.0.0.9 via FastEthernet0/3
2002-1-1 01:56:26       vers 2, CMD_RESPONSE, length 24
2002-1-1 01:56:26       10.1.1.0/24 via 0.0.0.0 metric 1
2002-1-1 01:56:43 RIP: ignored V1 packet from 10.1.1.1 (Illegal version).
//R2 路由器忽略了来自 10.1.1.1 的 v1 版本数据
2002-1-1 01:56:56 RIP: send to 224.0.0.9 via FastEthernet0/0
2002-1-1 01:56:56       vers 2, CMD_RESPONSE, length 24
2002-1-1 01:56:56       10.1.2.0/24 via 0.0.0.0 metric 1
2002-1-1 01:56:56 RIP: send to 224.0.0.9 via FastEthernet0/3
2002-1-1 01:56:56       vers 2, CMD_RESPONSE, length 24
2002-1-1 01:56:56       10.1.1.0/24 via 0.0.0.0 metric 1
```

以下是 R1 路由器中的 Debug 信息。

```
2002-1-1 03:40:14 RIP: send to 255.255.255.255 via FastEthernet0/0
```

```
2002-1-1 03:40:14          vers 1, CMD_RESPONSE, length 24
2002-1-1 03:40:14          10.1.3.0/0 via 0.0.0.0 metric 1
2002-1-1 03:40:14 RIP: send to 255.255.255.255 via FastEthernet0/1
2002-1-1 03:40:14          vers 1, CMD_RESPONSE, length 44
2002-1-1 03:40:14          10.1.1.0/0 via 0.0.0.0 metric 1
2002-1-1 03:40:14          10.1.2.0/0 via 0.0.0.0 metric 2
2002-1-1 03:40:27 RIP: recv RIP from 10.1.1.2 on FastEthernet0/0
2002-1-1 03:40:27          vers 2, CMD_RESPONSE, length 24
//R1 对 R2 的版本 2 信息是可以接收的，这从 R1 的路由表中也可以看到
2002-1-1 03:40:27          10.1.2.0/24 via 0.0.0.0 metric 1
```

R1 的路由表如下。

```
R1#sh ip route
Codes: C - connected, S - static, R - RIP, B - BGP, BC - BGP connected
       D - DEIGRP, DEX - external DEIGRP, O - OSPF, OIA - OSPF inter area
       ON1 - OSPF NSSA external type 1, ON2 - OSPF NSSA external type 2
       OE1 - OSPF external type 1, OE2 - OSPF external type 2
       DHCP - DHCP type
VRF ID: 0
C     10.1.1.0/24          is directly connected, FastEthernet0/0
R     10.1.2.0/24          [120,1] via 10.1.1.2(on FastEthernet0/0)
C     10.1.3.0/24          is directly connected, FastEthernet0/1
```

而 R2 的信息中却没有 10.1.3.0 的表项。

以上的操作说明了一个问题：RIPv1 是可以识别并采纳来自 RIPv2 的更新的，而 RIPv2 却不能识别版本 1 的信息，因此，此时可以在 RIPv2 的 R2 路由器中增加识别 RIPv1 更新的能力。

步骤 7：可以在 R2 的 F/0 端口中添加特殊的命令完成兼容性的配置，如下所示。

```
R2_config#interface fastEthernet 0/0
R2_config_f0/0#ip rip receive version 1
R2_config_f0/0#
```

此时，再次查看 R2 的路由表如下。

```
R2#sh ip route
Codes: C - connected, S - static, R - RIP, B - BGP, BC - BGP connected
       D - DEIGRP, DEX - external DEIGRP, O - OSPF, OIA - OSPF inter area
       ON1 - OSPF NSSA external type 1, ON2 - OSPF NSSA external type 2
       OE1 - OSPF external type 1, OE2 - OSPF external type 2
       DHCP - DHCP type
VRF ID: 0
C     10.1.1.0/24          is directly connected, FastEthernet0/0
C     10.1.2.0/24          is directly connected, FastEthernet0/3
R     10.1.3.0/24          [120,1] via 10.1.1.1(on FastEthernet0/0)
R2#
```

此时，从终端测试连通性，发现可以连通了。

任务扩展

1. 实现 RIP 被动接口的配置

RIP 协议由于使用广播更新并且常规 network 命令使整个网络充满了 RIP 报文，但某些终端网络并不需要这样的更新包，因此可以在设备中使用恰当的命令减少网络中的不必要的 RIP 报文，提升整网效率，如图 3-3-2 所示。

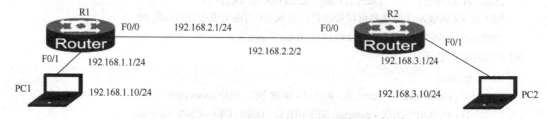

图 3-3-2　实现 RIP 被动接口的配置

要求：配置基础网络环境，配置 RIP 并使用 network 命令添加 192.168.2.0 网络，在 PC1 或者 PC2 上开启抓包软件捕获 RIP 报文，R1 和 R2 使用 network 命令增加与 PC 互连的网络，开启抓包软件捕获 RIP 报文，查看捕获结果，在两台路由器中使用 passive 命令停止 RIP 在此端口的发布更新，再次开启抓包软件捕获 RIP 报文，查看结果。

步骤 1：配置路由器 R1 的基础网络环境。

```
Router#config
Router_config#hostname R1
R1_config#interface fastEthernet 0/0
R1_config_f0/0#ip address 192.168.2.1 255.255.255.0
R1_config_f0/0#exit
R1_config#interface fastEthernet 0/1
R1_config_f0/1#ip address 192.168.1.1 255.255.255.0
R1_config_f0/1#exit
R1_config#router rip
R1_config_rip#network 192.168.1.0 255.255.255.0
R1_config_rip#network 192.168.2.0 255.255.255.0
R1_config_rip#ver 2
R1_config_rip#exit
R1_config#
```

步骤 2：配置路由器 R2 的基础网络环境。

```
Router#config
Router_config#hostname R2
R2_config#interface fastEthernet 0/0
R2_config_f0/0#ip address 192.168.2.2 255.255.255.0
R2_config_f0/0#exit
R2_config#interface fastEthernet 0/3
R2_config_f0/3#ip address 192.168.3.1 255.255.255.0
R2_config_f0/3#exit
```

```
R2_config#router rip
R2_config_rip#network 192.168.2.0 255.255.255.0
R2_config_rip#network 192.168.3.0 255.255.255.0
R2_config_rip#ver 2
R2_config_rip#exit
R2_config#
```

步骤 3：PC 中的抓包如图 3-3-3 所示。

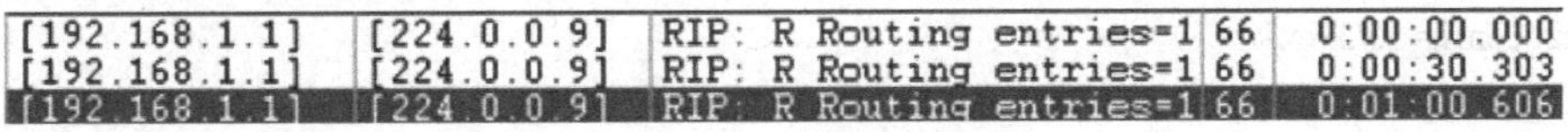

```
[192.168.1.1]    [224.0.0.9]    RIP: R Routing entries=1 66   0:00:00.000
[192.168.1.1]    [224.0.0.9]    RIP: R Routing entries=1 66   0:00:30.303
[192.168.1.1]    [224.0.0.9]    RIP: R Routing entries=1 66   0:01:00.606
```

图 3-3-3 抓包

在 PC1 中启动抓包过程，定义一个过滤器抓取 RIP 数据（注意，本任务抓取大约 3 个数据包，大约耗时 1.5min）。

从图 3-3-3 看出，在 PC1 中可以得到来自 192.168.1.1 的 RIP 报文，这是组播包，因为开启的是版本 2 的 RIP 协议，可以看到 R1 的路由器也在向没有路由器的 F0/1 接口发送数据，这是完全没有必要的，接下来可用命令消除这个网段的发布。

步骤 4：使用 passive 命令完成特定网段的消声处理。

```
R1_config#interface fastEthernet 0/1
R1_config_f0/1#ip rip passive
R1_config_f0/1#exit
```

此时，再次从 PC1 抓包，发现获取不到任何报文，查看 R2 的路由表，一切正常。

2. 实现 RIP v2 认证配置

安全问题在局域网中无处不在，路由协议的安全包括安全的更新过程，以确保路由信息的准确，在 RIP 协议中，版本 1 无法进行认证，而版本 2 可以提供更新路由器的认证，这样可以在某种程度上保证更新过程的准确可靠，如图 3-3-2 所示。

要求：配置基础网络环境，配置 RIP 协议，全部网段使用 network 命令，配置 RIP v2，启用认证。注意，配置 R1 的认证密钥与 R2 稍有不同或其中之一不配置。查看两台路由器的路由表，将认证密钥更新为一致，再次查看路由表。

步骤 1：配置路由器 R1 的基础网络环境。

```
Router_config#hostname R1
R1_config#interface fastEthernet 0/0
R1_config_f0/0#ip address 192.168.1.1 255.255.255.0
R1_config_f0/0#exit
R1_config#interface fastEthernet 0/1
R1_config_f0/1#ip address 192.168.2.1 255.255.255.0
R1_config_f0/1#exit
R1_config#router rip
R1_config_rip#network 192.168.1.0 255.255.255.0
R1_config_rip#network 192.168.2.0 255.255.255.0
R1_config_rip#version 2
R1_config_rip#exit
R1_config#
```

步骤 2：配置路由器 R2 的基础网络环境。

```
Router_config#hostname R2
R2_config#interface fastEthernet 0/0
R2_config_f0/0#ip add 192.168.1.2 255.255.255.0
R2_config_f0/0#exit
R2_config#interface fastEthernet 0/3
R2_config_f0/3#ip add 192.168.3.1 255.255.255.0
R2_config_f0/3#exit
R2_config#router rip
R2_config_rip#network 192.168.1.0 255.255.255.0
R2_config_rip#network 192.168.3.0 255.255.255.0
R2_config_rip#version 2
R2_config_rip#exit
R2_config#
```

步骤 3：R1 启动认证，使用 MD5 进行加密。

```
R1_config#interface fastEthernet 0/0
R1_config_f0/0#ip rip ?
  authentication         -- Set authentication mode
  message-digest-key     -- Set md5 authentication key and key-id
  passive                -- Only receive Update on the interface
  password               -- Set simple authentication password
  receive                -- Set receive version on the interface
  send                   -- Set send version on the interface
  split-horizon          -- Set split horizon on the interface
R1_config_f0/0#ip rip authentication ?
  message-digest         -- MD5 authentication
  simple                 -- Simple authentication
R1_config_f0/0#ip rip authentication message-digest ?
  <cr>
R1_config_f0/0#ip rip authentication message-digest
```

以上命令开启了 F0/0 端口上的 RIP MD5 认证。下面进入 MD5 密钥的配置过程。

```
R1_config_f0/0#ip rip message-digest-key ?
  <0-255>                -- key-ID
R1_config_f0/0#ip rip message-digest-key dcnu ?
ip rip message-digest-key dcnu ?
                        ^
Parameter invalid
```

以上的错误来自于"dcnu"参数的错误，更正方法就是使用数字，即

```
R1_config_f0/0#ip rip message-digest-key 1 ?
  md5                    -- Md5
R1_config_f0/0#ip rip message-digest-key 1 md5 ?
  WORD                   -- key(16 char)
R1_config_f0/0#ip rip message-digest-key 1 md5 1122334455667788
//此命令完成了 RIP 协议的 MD5 密钥的配置
```

此时没有配置 R2 的认证，查看路由表的结果如下。

```
R1#sh ip route
Codes: C - connected, S - static, R - RIP, B - BGP, BC - BGP connected
       D - DEIGRP, DEX - external DEIGRP, O - OSPF, OIA - OSPF inter area
       ON1 - OSPF NSSA external type 1, ON2 - OSPF NSSA external type 2
       OE1 - OSPF external type 1, OE2 - OSPF external type 2
       DHCP - DHCP type
VRF ID: 0
C       192.168.1.0/24          is directly connected, FastEthernet0/0
C       192.168.2.0/24          is directly connected, FastEthernet0/1
R1#
R2#sh ip route
Codes: C - connected, S - static, R - RIP, B - BGP, BC - BGP connected
       D - DEIGRP, DEX - external DEIGRP, O - OSPF, OIA - OSPF inter area
       ON1 - OSPF NSSA external type 1, ON2 - OSPF NSSA external type 2
       OE1 - OSPF external type 1, OE2 - OSPF external type 2
       DHCP - DHCP type
VRF ID: 0
C       192.168.1.0/24          is directly connected, FastEthernet0/0
C       192.168.3.0/24          is directly connected, FastEthernet0/3
R2#
```

开启 Debug 功能，查看结果如下。

```
-------------------------------R1-------------------------------
R1#2002-1-1 00:19:50 RIP: ignored V2 packet from 192.168.1.2 (Authentication failed)
2002-1-1 00:19:56 RIP: send to 224.0.0.9 via FastEthernet 0/0
2002-1-1 00:19:56          vers 2, CMD_RESPONSE, length 64
2002-1-1 00:19:56          192.168.2.0/24 via 0.0.0.0 metric 1
2002-1-1 00:19:56 RIP: send to 224.0.0.9 via FastEthernet 0/1
2002-1-1 00:19:56          vers 2, CMD_RESPONSE, length 24
2002-1-1 00:19:56          192.168.1.0/24 via 0.0.0.0 metric 1
-------------------------------R2-------------------------------
R2#debug ip rip packet
RIP protocol debugging is on
R2#2002-1-1 00:18:11 RIP: send to 224.0.0.9 via FastEthernet 0/0
2002-1-1 00:18:11          vers 2, CMD_RESPONSE, length 24
2002-1-1 00:18:11          192.168.3.0/24 via 0.0.0.0 metric 1
2002-1-1 00:18:11 RIP: send to 224.0.0.9 via FastEthernet 0/3
2002-1-1 00:18:11          vers 2, CMD_RESPONSE, length 24
2002-1-1 00:18:11          192.168.1.0/24 via 0.0.0.0 metric 1
2002-1-1 00:18:16 RIP: ignored V2 packet from 192.168.1.1 (Authentication failed)
```

上面的信息说明 R2 没有做认证，因此 R1 和 R2 都不处理来自对方的更新数据。
对 R2 配置验证，其过程如下。

```
R2_config#interface fastEthernet 0/0
R2_config_f0/0#ip rip authentication message-digest
```

R2_config_f0/0#ip rip message-digest-key 1 md5 1122334455667788
R2_config_f0/0#

此时，再次查看路由表，其结果如下。

```
-----------------------------R1-----------------------------
R1#sh ip route
Codes: C - connected, S - static, R - RIP, B - BGP, BC - BGP connected
       D - DEIGRP, DEX - external DEIGRP, O - OSPF, OIA - OSPF inter area
       ON1 - OSPF NSSA external type 1, ON2 - OSPF NSSA external type 2
       OE1 - OSPF external type 1, OE2 - OSPF external type 2
       DHCP - DHCP type
VRF ID: 0
    C        192.168.1.0/24        is directly connected, FastEthernet 0/0
    C        192.168.2.0/24        is directly connected, FastEthernet 0/1
    R        192.168.3.0/24        [120,1] via 192.168.1.2(on FastEthernet 0/0)
R1#
-----------------------------R2-----------------------------
R2#sh ip route
Codes: C - connected, S - static, R - RIP, B - BGP, BC - BGP connected
       D - DEIGRP, DEX - external DEIGRP, O - OSPF, OIA - OSPF inter area
       ON1 - OSPF NSSA external type 1, ON2 - OSPF NSSA external type 2
       OE1 - OSPF external type 1, OE2 - OSPF external type 2
       DHCP - DHCP type
VRF ID: 0
    C        192.168.1.0/24        is directly connected, FastEthernet 0/0
    R        192.168.2.0/24        [120,1] via 192.168.1.1(on FastEthernet 0/0)
    C        192.168.3.0/24        is directly connected, FastEthernet 0/3
R2#
```

任务四　实现 OSPF 单区域配置

需求分析

某公司的规模越来越大，路由器的数量也逐渐增多了，已经达到了 8 台。该公司的网络管理员经过考虑，发现原有的 RIP 协议已不再适合现有公司的应用了，因此，他决定在公司的路由器之间使用动态的 OSPF 协议，实现网络的互连。

方案设计

在大规模网络中，OSPF 作为链路状态路由协议的代表应用非常广泛，其具有无自环、收敛快的特点。使用中由于设备和软件版本不同，功能和配置方法将可能存在差异，请关注相应版本的使用说明。

所需设备如图 3-4-1 所示。

（1）DCR-2626 路由器 2 台。

（2）CR-V35MT 1 条。
（3）CR-V35FC 1 条。

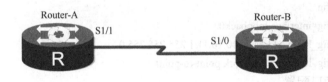

图 3-4-1　实现 OSPF 单区域配置

路由器配置信息见表 3-4-1。

表 3-4-1　路由器配置信息表

Router-A		Router-B	
接口	IP 地址	接口	IP 地址
S1/1	192.168.1.1/24	Loopback0	10.10.10.1/24
S1/0	192.168.1.2/24	Loopback0	10.10.11.1/24

知识准备

OSPF（Open Shortest Path First，开放式最短路径优先）是一个 IGP（Interior Gateway Protocol，内部网关协议），用于在单一自治系统内决策路由，是对链路状态路由协议的一种实现，隶属于内部网关协议。著名的 Dijkstra 算法被用来计算最短路径树。与 RIP 相比，OSPF 是链路状态协议，而 RIP 是距离矢量协议。不同厂商的管理距离不同，思科 OSPF 协议的管理距离是 110，华为 OSPF 协议的管理距离是 150。

1. OSPF 具体配置

启用 OSPF 动态路由协议：router ospf 进程号。
其中，进程号可以随意设置，只用于标识 OSPF 为本路由器内的一个进程。

2. 定义参与 OSPF 的子网。

定义 PSPF 的命令：network ip 子网号 通配符 area 区域号。
路由器将限制只能在相同区域内交换子网信息，不同区域间不交换路由信息。另外，
区域 0 为主干 OSPF 区域。不同区域交换路由信息必须经过区域 0。一般的，某一区域要接入 OSPF0 路由区域，则该区域必须至少有一台路由器为区域边缘路由器，即它既参与本区域路由，又参与区域 0 路由。
每个路由器的 OSPF 进程号可以不同，一个路由器可以有多个 OSPF 进程。OSPF 是无类路由协议，一定要加子网掩码，第一个区域必须是区域 0。

任务实现

步骤 1：路由器 A 环回接口的配置（其他接口配置请参见相关任务）。

Router-A_config#interface　loopback0　　　　　　　　　　//设置 loop back 口

```
Router-A_config_l0#ip address 10.10.10.1 255.255.255.0
Router-A_config_l0#ip ospf network point-to-point         //还原 loop back 口地址
```

步骤 2：路由器 B 环回接口的配置（其他接口配置请参见相关任务）。

```
Router-B#config
Router-B_config#interface loopback0
Router-B_config_l0#ip address   10.10.11.1 255.255.255.0
Router-B_config_l0#ip ospf network point-to-point
```

步骤 3：验证接口配置。

```
Router-B#sh interface loopback0
Loopback0 is up, line protocol is up
    Hardware is Loopback
    Interface address is 10.10.11.1/24
    MTU 1514 bytes, BW 8000000 kbit, DLY 500 usec
    Encapsulation LOOPBACK
```

步骤 4：路由器的 OSPF 配置。

Router-A 的配置如下。

```
Router-A_config#router ospf 1                //启动 OSPF 进程，进程号为 1
Router-A_config_ospf_1#network 10.10.10.0 255.255.255.0 area 0
//注意，要写子网掩码和区域号
Router-A_config_ospf_1#network 192.168.1.0 255.255.255.0 area 0
```

Router-B 的配置如下。

```
Router-B_config#router ospf 1
Router-B_config_ospf_1#network 10.10.11.0 255.255.255.0 area 0
Router-B_config_ospf_1#network 192.168.1.0 255.255.255.0 area 0
```

步骤 5：查看路由器 A 的路由表。

```
Router-A#sh ip route
Codes: C - connected, S - static, R - RIP, B - BGP, BC - BGP connected
       D - DEIGRP, DEX - external DEIGRP, O - OSPF, OIA - OSPF inter area
       ON1 - OSPF NSSA external type 1, ON2 - OSPF NSSA external type 2
       OE1 - OSPF external type 1, OE2 - OSPF external type 2
       DHCP - DHCP type
VRF ID: 0
C    10.10.10.0/24        is directly connected, Loopback0
O    10.10.11.1/32        [110,1600] via 192.168.1.2(on Serial1/1)
                          //注意，环回接口产生的是主机路由
C    192.168.1.0/24       is directly connected, Serial1/1
```

步骤 6：查看路由器 B 的路由表。

```
Router-B#show ip route
Codes: C - connected, S - static, R - RIP, B - BGP, BC - BGP connected
       D - DEIGRP, DEX - external DEIGRP, O - OSPF, OIA - OSPF inter area
       ON1 - OSPF NSSA external type 1, ON2 - OSPF NSSA external type 2
       OE1 - OSPF external type 1, OE2 - OSPF external type 2
       DHCP - DHCP type
```

VRF ID: 0			
O	10.10.10.1/32	[110,1601] via 192.168.1.1(on Serial1/0)	

//注意，管理距离为 110

C	10.10.11.0/24	is directly connected, Loopback0	
C	192.168.1.0/24	is directly connected, Serial1/0	

步骤 7：查看路由器 B 的进程信息。

Router-B#sh ip ospf 1　　　　　　　　　//显示该 OSPF 进程的信息

OSPF process: 1, Router ID: 192.168.2.1

Distance: intra-area 110,　inter-area 110,　external 150

SPF schedule delay 5 secs, Hold time between two SPFs 10 secs

SPFTV:11(1), TOs:24, SCHDs:27

All Rtrs support Demand-Circuit.

Number of areas is 1

AREA: 0

　Number of interface in this area is 2(UP: 3)

　Area authentication type:　None

　All Rtrs in this area support Demand-Circuit.

步骤 8：查看路由器 A 的进程信息。

Router-A#show ip ospf interface　　　　　//显示 OSPF 接口状态和类型

Serial1/1 is up, line protocol is up

　Internet Address: 192.168.1.1/24

　Nettype: Point-to-Point

　OSPF process is 2,　AREA: 0, Router ID: 192.168.1.1

　Cost: 1600, Transmit Delay is 1 sec, Priority 1

　Hello interval is 10, Dead timer is 40, Retransmit is 5

　OSPFINTFState is IPOINT_TO_POINT

　Neighbor Count is 1, Adjacent neighbor count is 1

　Adjacent with neighbor 192.168.1.2

Loopback0 is up, line protocol is up

　Internet Address: 10.10.10.1/24

　Nettype: Broadcast　　　　　　　　　//环回接口的网络类型默认为广播

　OSPF process is 2,　AREA: 0, Router ID: 192.168.1.1

　Cost: 1, Transmit Delay is 1 sec, Priority 1

　Hello interval is 10, Dead timer is 40, Retransmit is 5

　OSPFINTFState is ILOOPBACK

　Neighbor Count is 0, Adjacent neighbor count is 0

Router-A#sh ip ospf neighbor　　　　　　//显示 OSPF 邻居

OSPF process: 2

AREA: 0

Neighbor ID	Pri	State	DeadTime	Neighbor Addr	Interface
192.168.2.1	1	FULL/-	31	192.168.1.2	Serial1/1

步骤 9：修改环回接口的网络类型。

Router-A#conf

Router-A_config#interface　loopback0

Router-A_config_l0#ip ospf network point-to-point　　　　//将类型改为点到点

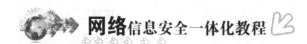

步骤10：查看路由器 A 的接口状态。

```
Router-A#sh ip ospf interface
Serial1/1 is up, line protocol is up
    Internet Address: 192.168.1.1/24
    Nettype: Point-to-Point
    OSPF process is 2,    AREA: 0, Router ID: 192.168.1.1
    Cost: 1600, Transmit Delay is 1 sec, Priority 1
    Hello interval is 10, Dead timer is 40, Retransmit is 5
    OSPFINTFState is IPOINT_TO_POINT
    Neighbor Count is 1, Adjacent neighbor count is 1
    Adjacent with neighbor 192.168.1.2
Loopback0 is up, line protocol is up
    Internet Address: 10.10.10.1/24
    Nettype: Point-to-Point
    OSPF process is 2,    AREA: 0, Router ID: 192.168.1.1
    Cost: 1, Transmit Delay is 1 sec, Priority 1
    Hello interval is 10, Dead timer is 40, Retransmit is 5
    OSPFINTFState is IPOINT_TO_POINT
    Neighbor Count is 0, Adjacent neighbor count is 0
```

步骤11：查看路由器 B 的路由表。

```
Router-B#sh ip route
Codes: C - connected, S - static, R - RIP, B - BGP, BC - BGP connected
       D - DEIGRP, DEX - external DEIGRP, O - OSPF, OIA - OSPF inter area
       ON1 - OSPF NSSA external type 1, ON2 - OSPF NSSA external type 2
       OE1 - OSPF external type 1, OE2 - OSPF external type 2
       DHCP - DHCP type
VRF ID: 0
O       10.10.10.0/24       [110,1600] via 192.168.1.1(on Serial1/0)
C       10.10.11.0/24       is directly connected, Loopback0
C       192.168.1.0/24      is directly connected, Serial1/0
```

任务五　实现 OSPF 多区域配置

♂ 需求分析

某公司的规模越来越大，路由器的数量也逐渐增多，已经达到了 20 台。该公司的网络管理员经过考虑，发现原有的 OSPF 单区域路由协议已不再适合现有公司的应用了，因此，他决定在公司的路由器之间使用动态的 OSPF 多区域配置，实现网络的互连。

♂ 方案设计

区域的概念是 OSPF 优于 RIP 的重要部分，它可以有效地提高路由的效率，缩减部分路由器的 OSPF 路由条目，降低路由收敛的复杂度，在区域边界上实现了路由的汇总、过滤、控制，大大提高了网络的稳定性。

所需设备如图 3-5-1 所示。

（1）DCR-2626 路由器 3 台。
（2）CR-V35FC 1 条。
（3）CR-V35MT 1 条。

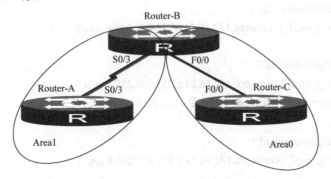

图 3-5-1 实现 OPSF 多区域配置

各路由器接口配置要求见表 3-5-1。

表 3-5-1 路由器接口配置表

Router-A		Router-B		Router-C	
接口	IP 地址	接口	IP 地址	接口	IP 地址
		Serial0/3	172.16.24.2/24	F0/0	172.16.25.2/24
Serial0/3	172.16.24.1/24	F0/0	172.16.25.1/24		
Loopback 0	10.10.10.1/24	Loopback 0	10.10.11.1/24	Loopback 0	10.10.12.1/24

知识准备

划分多区域可以减少链路状态数据库的大小，节约带宽；提高路由的效率；缩减部分路由器的 OSPF 路由条目，降低路由收敛的复杂度，对某些特定的链路状态广播，可以在区域边界上实现汇总、过滤、控制，从而实现全网互通；提高网络的稳定性；当某个区域的某条路由出现抖动时,可以减少受影响的范围。

区域 ID 长 32bit，其表示范围为 0～65535。当设置 Area 0 时，其区域 ID 为 0.0.0.0；当设置区域 Area 5 时，其区域 ID 为 0.0.0.5；当设置区域 Area 271 时，其区域 ID 为 0.0.1.15。使用区域时要注意以下几点。

（1）作为骨干区域的 Area 0 必须存在。
（2）所有区域，即使是端区，也必须和骨干区域相连。
（3）如果存在多个骨干区域，那么它们必须连续（逻辑上）。思考什么情况使用多骨干区域？优点是什么？
Area 0 只有一个最好，当有多个时其实是骨干区域分裂了，应使用虚链路连接。
（4）虚链路只能作为一种应急的临时策略。

任务实现

步骤 1：按照表 3-5-1 配置路由器名称、接口的 IP 地址，保证所有接口全部是 up 状态，测试连通性，此处略。

步骤 2：将 Router-A 和 Router-B 相应接口加入 Area 0。

```
Router-A:
Router-A_config#router ospf 1
Router-A_config_ospf_1#network 172.16.24.0 255.255.255.0 area 0
Router-B:
Router-B_config#router ospf 1
Router-B_config_ospf_1#network 172.16.24.0 255.255.255.0 area 0
```

步骤 3：将 Router-B、Router-C 相应接口加入 Area 1。

```
Router-B:
Router-B_config#router ospf 1
Router-B_config_ospf_1#network 172.16.25.0 255.255.255.0 area 1
Router-C:
Router-C_config#router ospf 1
Router-C_config_ospf_1# network 172.16.25.0 255.255.255.0 area 1
```

步骤 4：查看 Router-A、Router-C 上的 OSPF 路由表。

```
Router-A:
Router-A#show ip route
Codes: C - connected, S - static, R - RIP, B - BGP, BC - BGP connected
       D - DEIGRP, DEX - external DEIGRP, O - OSPF, OIA - OSPF inter area
       ON1 - OSPF NSSA external type 1, ON2 - OSPF NSSA external type 2
       OE1 - OSPF external type 1, OE2 - OSPF external type 2
       DHCP - DHCP type
VRF ID: 0
C        10.10.10.0/24         is directly connected, Loopback0
C        172.16.24.0/24        is directly connected, Serial0/2
O IA     172.16.25.0/24        [110,1601] via 172.16.24.2(on Serial0/2)
```

//提示学习到的是 OIA 区域间路由，OIA 的路由是通过 LSA3 来传播（后面的任务中会详解）的

```
Router-C:
Router-C# show ip route
Codes: C - connected, S - static, R - RIP, B - BGP, BC - BGP connected
       D - DEIGRP, DEX - external DEIGRP, O - OSPF, OIA - OSPF inter area
       ON1 - OSPF NSSA external type 1, ON2 - OSPF NSSA external type 2
       OE1 - OSPF external type 1, OE2 - OSPF external type 2
       DHCP - DHCP type
VRF ID: 0
C        10.10.12.0/24         is directly connected, Loopback0
O IA     172.16.24.0/24        [110,1601] via 172.16.25.1(on FastEthernet0/0)
C        172.16.25.0/24        is directly connected, FastEthernet0/0
```

//提示学习到的是 OIA 区域间路由，OIA 的路由是通过 LSA3 来传播（后面的任务中会详解）的

项目三
路由协议

任务六　实现 OSPF 虚链路配置

♂ 需求分析

在大规模网络中，通常通过划分区域来减少资源消耗，并将拓扑的变化本地化。由于实际环境的限制，不能物理地将其他区域环绕骨干区域，这样可以采用虚连接的方式逻辑地连接到骨干区域，使骨干区域自身也必须保持连通。

♂ 方案设计

如果非骨干区域无法和骨干区域直接连接，那么必须使用虚链路连接。虚链路只能作为一种应急的临时策略。

所需设备如图 3-6-1 所示。
（1）DCR-2611 3 台[版本为 1.3.3G (MIDDLE)]。
（2）CR-V35FC 1 条。
（3）CR-V35MT 1 条。

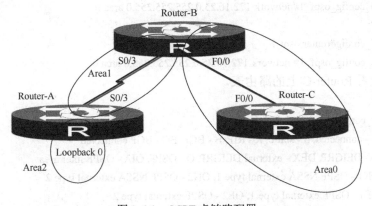

图 3-6-1　OSPF 虚链路配置

任务要求：按照拓扑图连接网络；按照要求配置路由器各接口地址，见表 3-6-1。

表 3-6-1　路由器接口配置

Router-A		Router-B		Router-C	
接口	IP 地址	接口	IP 地址	接口	IP 地址
S0/3（DCE）	172.16.24.1	S0/3(DTE)	172.16.24.2	F0/0	172.16.25.2
		F0/0	172.16.25.1		
Loopback 0	10.10.10.1	Loopback 0	172.16.25.1	Loopback 0	12.10.10.1

♂ 知识准备

虚链路必须配置在 ABR 上，在这个网络中 ABR 是 Router-A 和 Router-B。配置时是对端的 Router-ID，不是 IP 地址。虚链路被看作网络设计失败的一种补救手段，它不仅可以使没有和骨干区域直连的非骨干区域在逻辑上建立起一条链路，还可以连接两个分离的骨干区域。但

71 | PAGE

是由于虚链路的配置会造成日后维护和排错的困难，所以在进行网络设计的时候，不能将虚链路考虑进去。

任务实现

步骤 1：按照表 3-6-1 配置路由器名称、接口的 IP 地址，保证所有接口全部是 up 状态，测试连通性。

步骤 2：将 Router-A、Router-B 相应接口按照拓扑加入 Area 1、Area 2。

Router-A:
Router-A_config#router ospf 1
Router-A_config_ospf_1#network 10.10.10.0 255.255.255.0 area 2
Router-A_config_ospf_1#network 172.16.24.0 255.255.255.0 area 1
Router-B:
Router-B_config#router ospf 1
Router-B_config_ospf_1#network 172.16.24.0 255.255.255.0 area 1

步骤 3：将 Router-B、Router-C 相应接口按照拓扑加入 Area 0。

Router-B:
Router-B_config#router ospf 1
Router-B_config_ospf_1#network 172.16.25.0 255.255.255.0 area 0
Router-C:
Router-C_config#router ospf 1
Router-C_config_ospf_1# network 172.16.25.0 255.255.255.0 area 0

步骤 4：查看 Router-C 上的路由表。

Router-C:
Router-C_config_ospf_1#sh ip route
Codes: C - connected, S - static, R - RIP, B - BGP, BC - BGP connected
       D - DEIGRP, DEX - external DEIGRP, O - OSPF, OIA - OSPF inter area
       ON1 - OSPF NSSA external type 1, ON2 - OSPF NSSA external type 2
       OE1 - OSPF external type 1, OE2 - OSPF external type 2
       DHCP - DHCP type

VRF ID: 0
C   12.10.10.0/24   is directly connected, Loopback0
O IA   172.16.24.0/24   [110,1601] via 172.16.25.1(on FastEthernet0/0)
C   172.16.25.0/24   is directly connected, FastEthernet0/0
//只有 Area 1 传递来的 OSPF 路由，发现没有 Router-A 的 Loopback 接口路由

步骤 5：为 Router-A、Router-B 配置虚连接。

Router-A:
Router-A_config#router ospf 1
Router-A_config_ospf_1#area 1 virtual-link 11.10.10.1
Router-B:
Router-B_config#router ospf 1
Router-B_config_ospf_1#area 1 virtual-link 10.10.10.1
                            //注意，都是 Router-ID

步骤 6：在 Router-B 上查看虚链路状态。

Router-B:
Router-B_config_ospf_1#sh ip ospf virtual-link
Virtual Link Neighbor ID 10.10.10.1 (UP)
Run as Demand-Circuit
TransArea: 1, Cost is 1600
    Hello interval is 10, Dead timer is 40   Retransmit is 5
    INTF Adjacency state is IPOINT_TO_POINT
/*观察到已经建立了一条虚链路，虚链路在逻辑上等同于一条物理的按需链路，即只有在两端路由器的配置有变动的时候才进行更新*/

步骤7：查看Router-C上的路由表和OSPF数据库。

Router-C:
Router-C_config_ospf_1#sh ip route
Codes: C - connected, S - static, R - RIP, B - BGP, BC - BGP connected
       D - DEIGRP, DEX - external DEIGRP, O - OSPF, OIA - OSPF inter area
       ON1 - OSPF NSSA external type 1, ON2 - OSPF NSSA external type 2
       OE1 - OSPF external type 1, OE2 - OSPF external type 2
       DHCP - DHCP type
VRF ID: 0
O IA   10.10.10.1/32   [110,1602] via 172.16.25.1(on FastEthernet0/0)
C   12.10.10.0/24   is directly connected, Loopback0
O IA   172.16.24.0/24   [110,1601] via 172.16.25.1(on FastEthernet0/0)
O IA   172.16.24.2/32   [110,3201] via 172.16.25.1(on FastEthernet0/0)
C   172.16.25.0/24   is directly connected, FastEthernet0/0
/*已经学到了Router-A的Loopback接口路由。注意，其Metric值为1602，虚链路的Metric等同于所经过的全部链路开销之和，在这个网络中，Metric=1（Loopback）+到达Area 1的开销1061=1062*/

Router-C_config_ospf_1#show ip ospf database

                OSPF process: 1
              (Router ID: 12.10.10.1)
                AREA: 0
            Router Link States

Link ID	ADV Router	Age	SeqNum	Checksum	Link Count
10.10.10.1	10.10.10.1	1 (DNA)	0x80000003	0x0bab	1
11.10.10.1	11.10.10.1	435	0x80000005	0xf8fa	2
12.10.10.1	12.10.10.1	935	0x80000004	0x2b15	1

             Net Link States

Link ID	ADV Router	Age	SeqNum	Checksum
172.16.25.2	12.10.10.1	935	0x80000002	0x9070

          Summary Net Link States

Link ID	ADV Router	Age	SeqNum	Checksum
10.10.10.1	10.10.10.1	41 (DNA)	0x80000002	0x45b9
172.16.24.0	11.10.10.1	864	0x80000003	0xcd34
172.16.24.0	10.10.10.1	41 (DNA)	0x80000002	0xd82b
172.16.24.2	10.10.10.1	41 (DNA)	0x80000002	0xc43d

/*这里的（DNA）指DoNotAge，说明使用的是不老化的LSA，即虚链路是无需Hello包控制的*/

任务七 实现 OSPF 路由汇总

需求分析

假如公司路由区的 OSPF 区域中的子网是连续的,则区域边缘路由器向外传播给路由信息时,采用路由总结功能后,路由器会将所有连续的子网总结为一条路由传播给其他区域,则其他区域内的路由器看到这个区域的路由只有一条。这样可以节省路由时所需的网络带宽。

方案设计

在 OSPF 骨干区域中,一个区域的所有地址都会被通告进来。但是如果某个子网不稳定,那么在它每次改变状态的时候,都会引起 LSA 在整个网络中的泛洪。为了解决这个问题,可以对网络地址进行汇总。

所需设备如图 3-7-1 所示。

(1) DCR-2626 路由器 2 台。
(2) CR-V35FC 1 条。
(3) CR-V35MT 1 条。

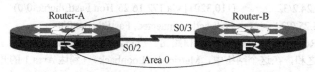

图 3-7-1 实现 OSPF 路由汇总

路由器各接口地址见表 3-7-1。

表 3-7-1 路由器配置表

Router-A		Router-B	
接口	IP 地址	接口	IP 地址
S0/2	172.16.24.1/24	S0/3	172.16.24.2/24
Loopback0	10.10.10.1/24	Loopback0	11.10.10.1/24

知识准备

在实际环境中,通常做精确的汇总;区域汇总就是区域之间的地址汇总,一般配置在 ABR 上;外部路由汇总就是一组外部路由通过重分布进入 OSPF,对这些外部路由进行汇总,一般配置在 ASBR 上。

任务实现

步骤 1:按照表 3-7-1 配置路由器名称、接口的 IP 地址,保证所有接口全部是 up 状态,测试连通性。

步骤 2:为 Router-A 配置 Loopback 接口 1~6。

Router-A:
Router-A_confi#int Loopback 1
Router-A_config_l1#ip add 1.1.0.1 255.255.255.0
Router-A_config_l1#int Loopback 2
Router-A_config_l2#ip add 1.2.0.1 255.255.255.0
Router-A_config_l2#int Loopback 3
Router-A_config_l3#ip add 1.3.0.1 255.255.255.0
Router-A_config_l3#int Loopback 4
Router-A_config_l4#ip add 1.4.0.1 255.255.255.0
Router-A_config_l4#int Loopback 5
Router-A_config_l5#ip add 1.5.0.1 255.255.255.0
Router-A_config_l5#int Loopback 6
Router-A_config_l6#ip add 1.6.0.1 255.255.255.0

步骤 3：为 Router-B 配置 Loopback 接口 1～6。

Router-B:
Router-B_config#int Loopback 1
Router-B_config_l1#ip add 2.1.0.1 255.255.255.0
Router-B_config_l1#int Loopback 2
Router-B_config_l2#ip add 2.2.0.1 255.255.255.0
Router-B_config_l2#int Loopback 3
Router-B_config_l3#ip add 2.3.0.1 255.255.255.0
Router-B_config_l3#int Loopback 4
Router-B_config_l4#ip add 2.4.0.1 255.255.255.0
Router-B_config_l4#int Loopback 5
Router-B_config_l5#ip add 2.5.0.1 255.255.255.0
Router-B_config_l5#int Loopback 6
Router-B_config_l6#ip add 2.6.0.1 255.255.255.0

步骤 4：将 Router-A、Router-B 相应接口按照拓扑加入 Area 0。

Router-A:
Router-A_config#router ospf 1
Router-A_config_ospf_1#network 172.16.24.0 255.255.255.0 area 0
Router-B:
Router-B_config#router ospf 1
Router-B_config_ospf_1#network 172.16.24.0 255.255.255.0 area 0

步骤 5：将 Router-B 配置为 ABR。

Router-B:
Router-B_config_ospf_1#network 2.1.0.0 255.255.0.0 area 1
Router-B_config_ospf_1#network 2.2.0.0 255.255.0.0 area 1
Router-B_config_ospf_1#network 2.3.0.0 255.255.0.0 area 1
Router-B_config_ospf_1#network 2.4.0.0 255.255.0.0 area 1
Router-B_config_ospf_1#network 2.5.0.0 255.255.0.0 area 1
Router-B_config_ospf_1#network 2.6.0.0 255.255.0.0 area 1

步骤 6：查看 Router-A 的 OSPF 路由表和数据库。

Router-A_config#sh ip route

```
Codes: C - connected, S - static, R - RIP, B - BGP, BC - BGP connected
       D - DEIGRP, DEX - external DEIGRP, O - OSPF, OIA - OSPF inter area
       ON1 - OSPF NSSA external type 1, ON2 - OSPF NSSA external type 2
       OE1 - OSPF external type 1, OE2 - OSPF external type 2
       DHCP - DHCP type
VRF ID: 0
    C       1.1.0.0/24              is directly connected, Loopback1
    C       1.2.0.0/24              is directly connected, Loopback2
    C       1.3.0.0/24              is directly connected, Loopback3
    C       1.4.0.0/24              is directly connected, Loopback4
    C       1.5.0.0/24              is directly connected, Loopback5
    C       1.6.0.0/24              is directly connected, Loopback6
    O IA    2.1.0.1/32              [110,1601] via 172.16.24.2(on Serial0/2)
    O IA    2.2.0.1/32              [110,1601] via 172.16.24.2(on Serial0/2)
    O IA    2.3.0.1/32              [110,1601] via 172.16.24.2(on Serial0/2)
    O IA    2.4.0.1/32              [110,1601] via 172.16.24.2(on Serial0/2)
    O IA    2.5.0.1/32              [110,1601] via 172.16.24.2(on Serial0/2)
    O IA    2.6.0.1/32              [110,1601] via 172.16.24.2(on Serial0/2)
    C       10.10.10.0/24           is directly connected, Loopback0
    C       172.16.24.0/24          is directly connected, Serial0/2
Router-A_config#show ip ospf database
-----------------------------------------------------------------------------
                         OSPF process: 1
                         (Router ID: 10.10.10.1)
                         AREA: 0
                     Router Link States
Link ID         ADV Router      Age     SeqNum      Checksum Link Count
10.10.10.1      10.10.10.1      255     0x80000003  0x8c03      2
11.10.10.1      11.10.10.1      162     0x80000004  0x7e0d      2
                     Summary Net Link States
Link ID         ADV Router      Age     SeqNum      Checksum
2.3.0.1         11.10.10.1      34      0x80000002  0x67af
2.4.0.1         11.10.10.1      34      0x80000002  0x5bba
2.5.0.1         11.10.10.1      34      0x80000002  0x4fc5
2.1.0.1         11.10.10.1      34      0x80000002  0x7f99
2.6.0.1         11.10.10.1      34      0x80000002  0x43d0
2.2.0.1         11.10.10.1      34      0x80000002  0x73a4
-----------------------------------------------------------------------------
/*可以发现不管是路由表还是数据库都比较大，为了解决这个问题，可在 ABR 上对红色部分配置域
内路由汇总*/
```

步骤 7：在 Router-B 上做域内路由汇总。

```
Router-B:
Router-B_config_ospf_1#area 1 range 2.0.0.0 255.248.0.0
                         //通过计算得出汇总的地址是 2.0.0.0/13
Router-B_config_ospf_1#exit
Router-B_config#ip route 2.0.0.0 255.248.0.0 null0
```

/*在进行区域汇总的时候，为了防止路由黑洞，一般会为这条汇总地址增加一条静态路由指向空接口（Null）*/

步骤 8：再次查看 Router-A 的路由表和数据库，比较汇总结果。

```
Router-A:
Router-A_config#sh ip route
Codes: C - connected, S - static, R - RIP, B - BGP, BC - BGP connected
       D - DEIGRP, DEX - external DEIGRP, O - OSPF, OIA - OSPF inter area
       ON1 - OSPF NSSA external type 1, ON2 - OSPF NSSA external type 2
       OE1 - OSPF external type 1, OE2 - OSPF external type 2
       DHCP - DHCP type
VRF ID: 0
    C    1.1.0.0/24         is directly connected, Loopback1
    C    1.2.0.0/24         is directly connected, Loopback2
    C    1.3.0.0/24         is directly connected, Loopback3
    C    1.4.0.0/24         is directly connected, Loopback4
    C    1.5.0.0/24         is directly connected, Loopback5
    C    1.6.0.0/24         is directly connected, Loopback6
    O IA 2.0.0.0/13         [110,1601] via 172.16.24.2(on Serial0/2)
    C    10.10.10.0/24      is directly connected, Loopback0
    C    172.16.24.0/24     is directly connected, Serial0/2
//观察到从原来的 6 条路由汇总成了一条 13 位的路由
Router-A_config#sh ip ospf database
------------------------------------------------------------------------
                    OSPF process: 1
                  (Router ID: 10.10.10.1)
                        AREA: 0
                    Router Link States
Link ID        ADV Router       Age       SeqNum       Checksum Link Count
10.10.10.1     10.10.10.1       936       0x80000003   0x8c03    2
11.10.10.1     11.10.10.1       843       0x80000004   0x7e0d    2
                  Summary Net Link States
Link ID        ADV Router       Age       SeqNum       Checksum
2.0.0.0        11.10.10.1       162       0x80000001   0x7ba7
------------------------------------------------------------------------
//LSA-3 也只剩下了一条，这大大减小了路由表和数据库的大小
```

任务八　实现 OSPF 认证配置

♂ 需求分析

某公司的路由器都使用了 OSPF 路由协议，现为了安全起见，想对相同区域的路由器启用身份验证功能，即只有经过身份验证的同一区域的路由器才能互相通告路由的信息，进行同步

更新。公司的网络管理员需要想办法实现此功能。

方案设计

与 RIP 相同，OSPF 也有认证机制，为了安全的原因，可以在相同 OSPF 区域的路由器上启用身份验证的功能，只有经过身份验证的同一区域的路由器才能互相通告路由信息。这样做可以增加网络安全性，对 OSPF 重新配置时，不同口令可以配置在新口令和旧口令的路由器上，防止它们在一个共享的公共广播网络中互相通信。

所需设备如图 3-8-1 所示。

（1）DCR-2611 路由器 2 台。
（2）CR-V35FC 1 条。
（3）CR-V35MT 1 条。

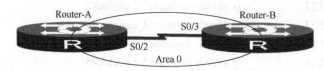

图 3-8-1 实现 OPSF 认证配置

路由器各接口地址见表 3-8-1。

表 3-8-1 路由器各接口地址

Router-A		Router-B	
接口	IP 地址	接口	IP 地址
S0/2	172.16.24.1/24	S0/3	172.16.24.2/24
Loopback0	10.10.10.1/24	Loopback0	11.10.10.1/24

知识准备

认证方式除加密之外，还有明文方式。区域验证是在 OSPF 路由进程下启用的，一旦启用，这台路由器所有属于这个区域的接口都将启用。接口验证是在接口下启用的，只影响路由器的一个接口。密码都在接口上配置，认证方式不同，会在不同的位置开启。

任务实现

步骤 1：按照表 3-8-1 配置路由器名称、接口的 IP 地址，保证所有接口全部是 up 状态，测试连通性。

步骤 2：将 Router-A、Router-B 相应接口按照拓扑加入 Area 0。

```
Router-A_config#router ospf 1
Router-A_config_ospf_1#network 172.16.24.0 255.255.255.0 area 0
Router-B_config#router ospf 1
Router-B_config_ospf_1#network 172.16.24.0 255.255.255.0 area 0
```

步骤 3：为 Router-A 接口配置 MD5 密文验证。

```
Router-A:
Router-A_config#interface S0/2
Router-A_config_s0/2#ip ospf message-digest-key 1 md5 DCNU
```

```
//采用 MD5 加密，密码为 DCNU
Router-A_config_s0/2#ip ospf authentication message-digest
//在 Router-A 上配置好后，启用 debug ip ospf packet，可以看到：
2002-1-1 00:02:39 OSPF: Send HELLO to 224.0.0.5 on Serial0/2
2002-1-1 00:02:39           HelloInt 10 Dead 40 Opt 0x2 Pri 1 len 44
2002-1-1 00:02:49 OSPF: Recv IP_SOCKET_RECV_PACKET message
2002-1-1 00:02:49 OSPF: Entering ospf_recv
2002-1-1 00:02:49 OSPF: Recv a packet from source: 172.16.24.2 dest 224.0.0.5
2002-1-1 00:02:49 OSPF: ERR recv PACKET, auth type not match
2002-1-1 00:02:49 OSPF: ERROR! events 21
```

这是因为 Router-A 发送了 key-id 为 1 的 key，但是 Router-B 上还没有配置验证，所以会出现验证类型不匹配的错误。

步骤 4：为 Router-B 接口配置 MD5 密文验证。

```
Router-B:
Router-B_config#interface S0/3
Router-B_config_s0/3#ip ospf message-digest-key 1 md5 DCNU     //定义 key 和密码
Router-B_config_s0/3#ip ospf authentication message-digest     //定义认证类型为 MD5
```

步骤 5：查看邻居关系。

```
Router-A:
Router-A#sh ip ospf neighbor
-----------------------------------------------------------------
 OSPF process: 1
 AREA: 0
Neighbor ID    Pri    State       DeadTime    Neighbor Addr     Interface
11.10.10.1     1      FULL/-      37          172.16.24.2       Serial0/2
-----------------------------------------------------------------
                                 //邻居关系已经建立
```

步骤 6：删除接口认证的配置，然后进行 OSPF 区域密文验证。

```
Router-A:
Router-A_config_ospf_1#area 0 authentication message-digest
Router-B:
Router-B_config_ospf_1#area 0 authentication message-digest
```

步骤 7：查看邻居关系。

```
Router-A:
Router-A#sh ip ospf neighbor
-----------------------------------------------------------------
 OSPF process: 1
 AREA: 0
Neighbor ID    Pri    State       DeadTime    Neighbor Addr     Interface
11.10.10.1     1      FULL/-      37          172.16.24.2       Serial0/2
-----------------------------------------------------------------
                                 //邻居关系已经建立
```

认证考核

选择题

1. 在一个运行 OSPF 的自治系统之内，（　　）。
   A．骨干区域自身也必须是连通的
   B．非骨干区域自身也必须是连通的
   C．必须存在一个骨干区域（区域号为0）
   D．非骨干区域与骨干区域必须直接相连或逻辑上相连

2. 下列关于 OSPF 协议的说法正确的是（　　）。
   A．OSPF 支持基于接口的报文验证
   B．OSPF 支持到同一目的地址的多条等值路由
   C．OSPF 是一个基于链路状态算法的边界网关路由协议
   D．OSPF 发现的路由可以根据不同的类型而有不同的优先级

3. 下列静态路由配置正确的是（　　）。
   A．ip route 129.1.0.0 16 serial 0
   B．ip route 10.0.0.2 16 129.1.0.0
   C．ip route 129.1.0.0 16 10.0.0.2
   D．ip route 129.1.0.0 255.255.0.0 10.0.0.2

4. 以下不属于动态路由协议的是（　　）。
   A．RIP　　　　B．ICMP　　　　C．IS-IS　　　　D．OSPF

5. 三种路由协议 RIP、OSPF、BGP 和静态路由各自得到了一条到达目标网络的路径，默认情况下，最终选定（　　）作为最优路由。
   A．RIP　　　　B．OSPF　　　　C．BGP　　　　D．静态路由

6. 路由环问题会引起（　　）。
   A．慢收敛　　　B．广播风暴　　　C．路由器重启　　　D．路由不一致

7. 以下路由表项需要由网络管理员手动配置的是（　　）。
   A．静态路由　　B．直接路由　　C．动态路由　　D．以上说法都不正确

8. 关于 RIP 协议，下列说法正确的有（　　）。
   A．RIP 协议是一种 IGP
   B．RIP 协议是一种 EGP
   C．RIP 协议是一种距离矢量路由协议
   D．RIP 协议是一种链路状态路由协议

9. RIP 协议基于（　　）实现。
   A．UDP　　　　B．TCP　　　　C．ICMP　　　　D．Raw IP

10. RIP 协议的路由项在（　　）内没有更新会变为不可达。
    A．90s　　　　B．120s　　　　C．180s　　　　D．240s

11. 解决路由环路的方法有（　　）。
    A．水平分割　　B．抑制时间　　C．毒性逆转　　D．触发更新

项目四 路由重分布与策略路由

教学背景

路由重分布将一种路由选择协议获悉的网络告知另一种路由选择协议，以便网络中每个工作站都能到达其他的任意工作站，这一过程被称为路由重分布。一般来说，一个组织或一个跨国公司很少只使用一种路由协议，而如果一个公司同时运行了多种路由协议，或者一个公司和另外一个公司合并的时候两个公司使用的路由协议不一样，此时该怎么办呢？此时必须采取一种方式来将一种路由协议的信息发布到另外一种路由协议中，这样，重分布的技术就诞生了。

策略路由是一种比基于目标网络进行路由更加灵活的数据包路由转发机制。应用了策略路由，路由器将通过路由图决定如何对需要路由的数据包进行处理，路由图决定了一个数据包的下一跳转发路由器。

任务一 静态路由和 RIP 路由的重分布

需求分析

某公司兼并了一个公司，现在需要在两个公司之间建立网络连接，实现全网互通，但是由于两个公司所采用的网络协议不同，因此建立连接的工作出现了困难。

方案设计

网络环境多种多样，很多时候需要在不同的协议之间对路由进行重分布，以使相关网络信息传递到需要的网络中。公司原来所采用的网络协议为 RIP，而新兼并的公司采用的路由协议是静态路由协议。为了成功完成网络连接，工程师采用静态路由和 RIP 重分布的技术来解决问题。

所需设备如图 4-1-1 所示。
（1）DCR-2626 路由器 3 台。
（2）PC 2 台。
（3）网线若干。

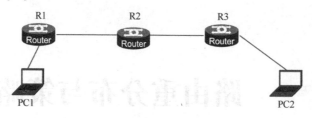

图 4-1-1 静态路由和 RIP 路由的重分布

任务要求：路由器接口配置信息，见表 4-1-1。配置基础网络环境，配置 R2 使用静态路由到达 192.168.1.0 网络，通过 RIP 协议学习到 192.168.3.0 网络。R2 中的 network 只增加 192.168.4.0 网络。R1 使用默认路由 192.168.2.2 到达其他远程网络。R3 使用 RIP 协议与 R2 交互学习网络信息。在没有做任何重分布配置时，查看 3 台路由器的路由表。在 R2 中做静态路由的重分布，查看 R3 的路由表如何？在 R2 中增加直连路由的重分布，查看 R3 的路由表有何变化？

表 4-1-1 路由器接口配置信息

	R1	R2	R3
F0/0	192.168.2.1	192.168.2.2	192.168.4.1
F0/1		192.168.4.2	
F0/3	192.168.1.1		192.168.3.1

知识准备

在大型的企业中，可能在同一网络内使用多种路由协议，为了实现多种路由协议的协同工作，路由器可以使用路由重分布将其学习到的一种路由协议的路由通过另一种路由协议广播出去，这样网络的所有部分都可以连通了。为了实现重分布，路由器必须同时运行多种路由协议，这样，每种路由协议才可以取得路由表中的所有或部分其他协议的路由来进行广播。

redistribute 命令可以用来实现路由重分布，它既可以重分布所有路由，又可以根据匹配的条件，选择某些路由进行重分布，此外，该命令还支持某些参数的设置，如设置 metric。

完整的 redistribute 命令格式如下。

redistribute protocol [process-id] [level-1 | level-1-2 | level-2] [as-number] [metric metric-value] [metric-type type-value] [match {internal | external 1 | external 2}] [tag tag-value] [route-map map-tag] [subnets]

redistribute 命令标明了重分布路由的来源，而 router 命令标明了广播路由的进程。例如，命令 redistribute eigrp 1 表示路由器对来自 EIGRP 进程 1 的路由进行重分布；如果该命令在 router rip 下，则该路由被重分布为 RIP 路由，这样其他 RIP 路由器即可看到来自 EIGRP AS 1 的路由。

在路由器上配置多路由协议间的重分布，如将路由协议 A 重分布到路由协议 B 中，要先进入路由协议 B 的路由模式下，再执行 redistribute 命令进行重分布的操作，并配置相应的路由选路参数。

一般做多路由协议间的重分布时要做双向的，即将路由协议 A 重分布到路由协议 B 后，再执行路由协议 B 到路由协议 A 的重分布，或配置单向的重分布后，再添加一条指向到对方的默认路由（一般用于外部路由协议间，如配置 BGP 时）。路由重分布，即将一种路由协议中的路由条目转换为另一种路由协议的路由条目，实现多路由环境下的网络互通。

任务实现

步骤 1：配置路由器 R1 的基础网络环境。

```
Router_config#hostname R1
R1_config#interface fastEthernet 0/0
R1_config_f0/0#ip address 192.168.2.1 255.255.255.0
R1_config_f0/0#exit
R1_config#interface fastEthernet 0/3
R1_config_f0/3#ip add 192.168.1.1 255.255.255.0
R1_config_f0/3#exit
R1_config#
```

步骤 2：配置路由器 R2 的基础网络环境。

```
Router_config#hostname R2
R2_config#interface fastEthernet 0/0
R2_config_f0/0#ip address 192.168.2.2 255.255.255.0
R2_config_f0/0#exit
R2_config#interface fastEthernet 0/1
R2_config_f0/1#ip address 192.168.4.2 255.255.255.0
R2_config_f0/1#exit
R2_config#
```

步骤 3：配置路由器 R3 的基础网络环境。

```
Router_config#hostname R3
R3_config#interface fastEthernet 0/0
R3_config_f0/0#ip address 192.168.4.1 255.255.255.0
R3_config_f0/0#exit
R3_config#interface fastEthernet 0/3
R3_config_f0/3#ip address 192.168.3.1 255.255.255.0
R3_config_f0/3#exit
R3_config#
```

步骤 4：测试 R1 链路的连通性。

```
R1_config#ping 192.168.1.10
PING 192.168.1.10 (192.168.1.10): 56 data bytes
!!!!!
--- 192.168.1.10 ping statistics ---
5 packets transmitted, 5 packets received, 0% packet loss
round-trip min/avg/max = 0/0/0 ms
R1_config#ping 192.168.2.2
PING 192.168.2.2 (192.168.2.2): 56 data bytes
!!!!!
--- 192.168.2.2 ping statistics ---
5 packets transmitted, 5 packets received, 0% packet loss
round-trip min/avg/max = 0/0/0 ms
R1_config#
```

步骤 5：测试 R2 链路的连通性。

```
R2#ping 192.168.2.1
PING 192.168.2.1 (192.168.2.1): 56 data bytes
!!!!!
--- 192.168.2.1 ping statistics ---
5 packets transmitted, 5 packets received, 0% packet loss
round-trip min/avg/max = 0/0/0 ms
R2#ping 192.168.4.1
PING 192.168.4.1 (192.168.4.1): 56 data bytes
!!!!!
--- 192.168.4.1 ping statistics ---
5 packets transmitted, 5 packets received, 0% packet loss
round-trip min/avg/max = 0/0/0 ms
R2#
```

步骤 6：测试 R3 链路的连通性。

```
R3_config#ping 192.168.4.2
PING 192.168.4.2 (192.168.4.2): 56 data bytes
!!!!!
--- 192.168.4.2 ping statistics ---
5 packets transmitted, 5 packets received, 0% packet loss
round-trip min/avg/max = 0/0/0 ms
R3_config#
R3_config#ping 192.168.3.10
PING 192.168.3.10 (192.168.3.10): 56 data bytes
!!!!!
--- 192.168.3.10 ping statistics ---
5 packets transmitted, 5 packets received, 0% packet loss
round-trip min/avg/max = 0/0/0 ms
R3_config#
```

步骤 7：配置 R1 路由的环境，R1 主要使用静态路由。

```
R1_config#ip route 192.168.4.0 255.255.255.0 192.168.2.2
R1_config#ip route 192.168.3.0 255.255.255.0 192.168.2.2
```

也可以添加如下所示的默认路由：

```
R1_config#ip route 0.0.0.0 0.0.0.0 192.168.2.2
R1_config#
```

以上两种方法在本任务中效果是一样的。

步骤 8：配置 R2 路由，在 f0/0 上使用静态路由，而在 f0/1 上使用 RIP 协议。

```
R2_config#ip route 192.168.1.0 255.255.255.0 192.168.2.1
R2_config#exit
R2_config#router rip
R2_config_rip#network 192.168.4.0 255.255.255.0
R2_config_rip#ver 2
R2_config_rip#exit
R2_config#exit
```

步骤 9：配置 R3 路由器，使用 RIP 协议完成路由环境搭建。

R3_config#router rip
R3_config_rip#network 192.168.4.0 255.255.255.0
R3_config_rip#network 192.168.3.0 255.255.255.0
R3_config_rip#ver 2
R3_config_rip#exit

步骤 10：查看 R1 的路由表。

R1_config#sh ip route
Codes: C - connected, S - static, R - RIP, B - BGP, BC - BGP connected
　　　　D - DEIGRP, DEX - external DEIGRP, O - OSPF, OIA - OSPF inter area
　　　　ON1 - OSPF NSSA external type 1, ON2 - OSPF NSSA external type 2
　　　　OE1 - OSPF external type 1, OE2 - OSPF external type 2
　　　　DHCP - DHCP type

VRF ID: 0
S　　　0.0.0.0/0　　　　　　[1,0] via 192.168.2.2(on FastEthernet0/0)
C　　　192.168.1.0/24　　　is directly connected, FastEthernet0/3
C　　　192.168.2.0/24　　　is directly connected, FastEthernet0/0
S　　　192.168.3.0/24　　　[1,0] via 192.168.2.2(on FastEthernet0/0)
S　　　192.168.4.0/24　　　[1,0] via 192.168.2.2(on FastEthernet0/0)
R1_config#

上面的路由表是编者既添加静态路由又添加默认路由的情况，如果只添加默认路由，则没有后两条静态路由；如果只添加静态路由，则没有最上面的默认路由。

步骤 11：查看 R2 的路由表。

R2_config#sh ip route
Codes: C - connected, S - static, R - RIP, B - BGP, BC - BGP connected
　　　　D - DEIGRP, DEX - external DEIGRP, O - OSPF, OIA - OSPF inter area
　　　　ON1 - OSPF NSSA external type 1, ON2 - OSPF NSSA external type 2
　　　　OE1 - OSPF external type 1, OE2 - OSPF external type 2
　　　　DHCP - DHCP type

VRF ID: 0
S　　　192.168.1.0/24　　　[1,0] via 192.168.2.1(on FastEthernet0/0)
C　　　192.168.2.0/24　　　is directly connected, FastEthernet0/0
R　　　192.168.3.0/24　　　[120,1] via 192.168.4.1(on FastEthernet0/1)
C　　　192.168.4.0/24　　　is directly connected, FastEthernet0/1
R2_config#

步骤 12：查看 R3 的路由表。

R3_config#sh ip route
Codes: C - connected, S - static, R - RIP, B - BGP, BC - BGP connected
　　　　D - DEIGRP, DEX - external DEIGRP, O - OSPF, OIA - OSPF inter area
　　　　ON1 - OSPF NSSA external type 1, ON2 - OSPF NSSA external type 2
　　　　OE1 - OSPF external type 1, OE2 - OSPF external type 2
　　　　DHCP - DHCP type

VRF ID: 0
C　　　192.168.3.0/24　　　is directly connected, FastEthernet0/3

```
C       192.168.4.0/24          is directly connected, FastEthernet0/0
R3_config#
```

分析结果，可以知道，此时只有 R1 和 R2 的路由表是完整的，因为 R1 使用静态路由，而 R2 与 R3 建立了完整的 RIP 更新环境，对于 R3 来说，由于 R2 没有把左侧网络的情况添加到路由进程中，因此 R3 什么新消息都无法得到。

步骤 13：添加一个直连路由的重分布命令给 R2。

```
R2_config#router rip
R2_config_rip#redistribute connect
R2_config_rip#exit
```

步骤 14：在 R3 中查看路由表。

```
R3_config#sh ip route
Codes: C - connected, S - static, R - RIP, B - BGP, BC - BGP connected
       D - DEIGRP, DEX - external DEIGRP, O - OSPF, OIA - OSPF inter area
       ON1 - OSPF NSSA external type 1, ON2 - OSPF NSSA external type 2
       OE1 - OSPF external type 1, OE2 - OSPF external type 2
       DHCP - DHCP type

VRF ID: 0
    R       192.168.2.0/24          [120,1] via 192.168.4.2(on FastEthernet0/0)
    C       192.168.3.0/24          is directly connected, FastEthernet0/3
    C       192.168.4.0/24          is directly connected, FastEthernet0/0
R3_config#
```

可以看出，经过直连路由的重分布，R2 将其直连网段 192.168.2.0 分布给了 R3，但没有将静态路由（S 标识的那些静态路由）分布给 R3。

步骤 15：如果 R3 需要获得完整的路由，则需要在 R2 中添加静态路由的重分布。

```
R2_config#router rip
R2_config_rip#redistribute static
R2_config_rip#exit
```

步骤 16：再次从 R3 中查看结果。

```
R3_config#sh ip route
Codes: C - connected, S - static, R - RIP, B - BGP, BC - BGP connected
       D - DEIGRP, DEX - external DEIGRP, O - OSPF, OIA - OSPF inter area
       ON1 - OSPF NSSA external type 1, ON2 - OSPF NSSA external type 2
       OE1 - OSPF external type 1, OE2 - OSPF external type 2
       DHCP - DHCP type

VRF ID: 0
    R       192.168.1.0/24          [120,1] via 192.168.4.2(on FastEthernet0/0)
    R       192.168.2.0/24          [120,1] via 192.168.4.2(on FastEthernet0/0)
    C       192.168.3.0/24          is directly connected, FastEthernet0/3
    C       192.168.4.0/24          is directly connected, FastEthernet0/0
R3_config#
```

至此，通过对静态路由区域和直连路由的重分布，将 RIP 环境完整地搭建了起来。

任务二 RIP 和 OSPF 的重分布

需求分析

某公司兼并了一个公司，现在需要在两个公司之间建立网络连接，但是由于两个公司采用的网络协议不同，使得建立连接的工作出现了困难。

方案设计

RIP 和 OSPF 协议是目前使用较频繁的路由协议，两种协议交接的场合也很多见，它们之间的重分布是比较常见的配置。公司原来所采用的网络协议为 OSPF，而新兼并过来的公司采用的路由协议是 RIP。为了成功完成网络的连接，工程师采用 RIP 和 OSPF 重分布技术来解决此问题。

所需设备如图 4-2-1 所示。
（1）DCR-2626 路由器 3 台。
（2）PC 2 台。
（3）网线若干。

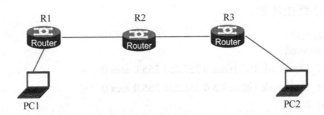

图 4-2-1 RIP 和 OSPF 的重分布

任务要求：配置基础环境，如表 4-2-1 所示。配置 R1 和 R2 之间使用 RIP 协议学习路由信息，R2 和 R3 之间使用 OSPF 协议。R1 配置 RIP 协议，使用两条 network 命令包括两个直连网络；R2 配置 RIP 使用一条 network 命令包括 192.168.2.0 网络，配置 OSPF 使用一条 network 命令包括 192.168.4.0 网络；R3 配置 OSPF 协议两条 network 命令包括所有直连网络。在 R2 中配置 RIP 到 OSPF 的重分布，再配置 OSPF 到 RIP 的重分布，查看路由表。

表 4-2-1 路由器接口配置信息

	R1	R2	R3
F0/0	192.168.2.1	192.168.2.2	192.168.4.1
F0/1		192.168.4.2	
F0/3	192.168.1.1		192.168.3.1

任务实现

步骤 1：配置基础网络环境，此处略。
步骤 2：配置 R1 路由环境。
R1 配置纯 RIP 环境，即使用 network 命令指定相邻网段进入 RIP 进程，其过程如下：

R1#config

```
R1_config#router rip
R1_config_rip#network 192.168.1.0 255.255.255.0
R1_config_rip#network 192.168.2.0 255.255.255.0
R1_config_rip#version 2
R1_config_rip#exit
R1_config#exit
R1#wr
```

步骤 3：配置 R2 路由环境。

R2 环境相对复杂一些，配置 F0/0 端口网段使用 RIP 协议，配置 F0/1 端口网段使用 OSPF 协议，其过程如下：

```
R2_config#router rip
R2_config_rip#version 2
R2_config_rip#network 192.168.2.0 255.255.255.0
R2_config_rip#exit
R2_config#router ospf 1
R2_config_ospf_1#network 192.168.4.0 255.255.255.0 area 0
R2_config_ospf_1#exit
R2_config#
```

步骤 4：配置 R3 路由环境。

```
R3#config
R3_config#router ospf 1
R3_config_ospf_1#network 192.168.4.0 255.255.255.0 area 0
R3_config_ospf_1#network 192.168.3.0 255.255.255.0 area 0
R3_config_ospf_1#exit
R3_config#
```

步骤 5：查看 R1 的路由表。

```
R1_config#sh ip route
Codes: C - connected, S - static, R - RIP, B - BGP, BC - BGP connected
       D - DEIGRP, DEX - external DEIGRP, O - OSPF, OIA - OSPF inter area
       ON1 - OSPF NSSA external type 1, ON2 - OSPF NSSA external type 2
       OE1 - OSPF external type 1, OE2 - OSPF external type 2
       DHCP - DHCP type

VRF ID: 0
C       192.168.1.0/24       is directly connected, FastEthernet0/3
C       192.168.2.0/24       is directly connected, FastEthernet0/0
R1_config#
```

步骤 6：查看 R2 的路由表。

```
R2_config#sh ip route
Codes: C - connected, S - static, R - RIP, B - BGP, BC - BGP connected
       D - DEIGRP, DEX - external DEIGRP, O - OSPF, OIA - OSPF inter area
       ON1 - OSPF NSSA external type 1, ON2 - OSPF NSSA external type 2
       OE1 - OSPF external type 1, OE2 - OSPF external type 2
       DHCP - DHCP type
```

```
VRF ID: 0
R       192.168.1.0/24      [120,1] via 192.168.2.1(on FastEthernet0/0)
C       192.168.2.0/24      is directly connected, FastEthernet0/0
O       192.168.3.0/24      [110,2] via 192.168.4.1(on FastEthernet0/1)
C       192.168.4.0/24      is directly connected, FastEthernet0/1
R2_config#
```

步骤 7：查看 R3 的路由表。

```
R3_config#sh ip route
Codes: C - connected, S - static, R - RIP, B - BGP, BC - BGP connected
       D - DEIGRP, DEX - external DEIGRP, O - OSPF, OIA - OSPF inter area
       ON1 - OSPF NSSA external type 1, ON2 - OSPF NSSA external type 2
       OE1 - OSPF external type 1, OE2 - OSPF external type 2
       DHCP - DHCP type
VRF ID: 0
C       192.168.3.0/24      is directly connected, FastEthernet0/3
C       192.168.4.0/24      is directly connected, FastEthernet0/0
R3_config#
```

从路由表上可以发现，只有 R2 的路由表是完整的，R1 和 R3 都因为 R2 没有将对方的信息进行传递而得不到远端网络的消息，所以问题的关键是 R2。

步骤 8：在 R2 中启用动态路由的重分布过程，仅将 RIP 协议重分布到 OSPF 协议。

```
R2_config#router ospf 1
R2_config_ospf_1#redistribute rip
R2_config_ospf_1#exit
R2_config#
```

步骤 9：此时查看 R3 的路由表。

```
R3#sh ip route
Codes: C - connected, S - static, R - RIP, B - BGP, BC - BGP connected
       D - DEIGRP, DEX - external DEIGRP, O - OSPF, OIA - OSPF inter area
       ON1 - OSPF NSSA external type 1, ON2 - OSPF NSSA external type 2
       OE1 - OSPF external type 1, OE2 - OSPF external type 2
       DHCP - DHCP type
VRF ID: 0
O E2    192.168.1.0/24      [150,100] via 192.168.4.2(on FastEthernet0/0)
C       192.168.3.0/24      is directly connected, FastEthernet0/3
C       192.168.4.0/24      is directly connected, FastEthernet0/0
R3#
```

从路由表中可以看到，它学习到了一条 OSPF 自治系统外部路由（因为是从 RIP 协议注入的），其默认的初始度量值是 100。但同时也观察到，这个路由表依然是不完整的，因为它没有学习到 R2 的直连网络 192.168.2.0。

步骤 10：在 R2 中将直连路由在 OSPF 进程中重分布一下。

```
R2_config#router ospf 1
R2_config_ospf_1#redistribute connect
```

步骤 11：再次查看 R3 的路由表。

```
R3#sh ip route
Codes: C - connected, S - static, R - RIP, B - BGP, BC - BGP connected
       D - DEIGRP, DEX - external DEIGRP, O - OSPF, OIA - OSPF inter area
       ON1 - OSPF NSSA external type 1, ON2 - OSPF NSSA external type 2
       OE1 - OSPF external type 1, OE2 - OSPF external type 2
       DHCP - DHCP type
VRF ID: 0
O E2    192.168.1.0/24      [150,100] via 192.168.4.2(on FastEthernet0/0)
O E2    192.168.2.0/24      [150,100] via 192.168.4.2(on FastEthernet0/0)
C       192.168.3.0/24      is directly connected, FastEthernet0/3
C       192.168.4.0/24      is directly connected, FastEthernet0/0
```

R3 已经完整了，再来观察 R1 有没有变化。观察到依然没有变化，它的路由表完整性是依赖 R2 通过 RIP 协议传递的，而 RIP 协议传递的消息并没有包含其他远端网络。

步骤 12：在 R2 中做 RIP 进程的重分布。

```
R2_config#router rip
R2_config_rip#redistributeospf 1
R2_config_rip#redistribute connect
R2_config_rip#exit
```

步骤 13：再次查看 R1 的路由表。

```
R1_config#sh ip route
Codes: C - connected, S - static, R - RIP, B - BGP, BC - BGP connected
       D - DEIGRP, DEX - external DEIGRP, O - OSPF, OIA - OSPF inter area
       ON1 - OSPF NSSA external type 1, ON2 - OSPF NSSA external type 2
       OE1 - OSPF external type 1, OE2 - OSPF external type 2
       DHCP - DHCP type
VRF ID: 0
C       192.168.1.0/24      is directly connected, FastEthernet0/3
C       192.168.2.0/24      is directly connected, FastEthernet0/0
R       192.168.3.0/24      [120,1] via 192.168.2.2(on FastEthernet0/0)
R       192.168.4.0/24      [120,1] via 192.168.2.2(on FastEthernet0/0)
R1_config#
```

此时，从终端测试连通性，发现可以连通，本任务完成。

任务三　基于源地址的策略路由

♂ 需求分析

由于公司业务地域范围较大，主要分成南方客户和北方客户；PC1 主要负责北方的业务，PC2 主要负责南方的业务。公司网络出口有两条专线，分别是网通和电信。公司的网络出口多了，本来应该在对外访问的时候速度更快。但是从运行一段时间的情况来看，网络速度并没有绝对加快。

方案设计

从局域网去往广域网的流量有时需要进行分流,即区分不同用户又进行了负载分担,有时这种目标是通过对不同的源地址进行区分来完成的。

所需设备如图 4-3-1 所示。

(1) DCR-2626 路由器 3 台。
(2) PC 2 台。
(3) 网线若干。

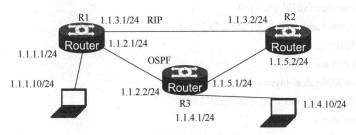

图 4-3-1 基于源地址的策略路由

任务要求:配置基础网络环境,全网使用 OSPF 单区域完成路由的连通,在 R3 中使用策略路由,使来自 1.1.4.10 的源地址去往外网的路由从 1.1.2.1 通过,而来自 1.1.4.20 的源地址的数据从 1.1.5.2 的路径通过;跟踪从 1.1.4.10 去往 1.1.1.10 的数据路由;将 1.1.4.10 地址改为 1.1.4.20,再次跟踪路由。

知识准备

策略路由是一种比基于目标网络进行路由更加灵活的数据包路由转发机制。应用了策略路由,路由器将通过路由图决定如何对需要路由的数据包进行处理,路由图决定了一个数据包的下一跳转发路由器。

基于策略的路由为网络管理员提供了比传统路由协议对报文的转发和存储更强的控制能力,传统上,路由器应用从路由协议派生出来的路由表,根据目的地址进行报文的转发。

基于策略的路由比传统路由强,使用更灵活,它使网络管理员无法根据目的地址,而根据报文大小、应用或 IP 源地址来选择转发路径。策略可以定义为通过多路由器的负载平衡或根据总流量在各线上进行转发的服务质量。而在当今高性能的网络中,这种选择的自由性是很需要的。

策略路由提供了这样一种机制:根据网络管理员制定的标准来进行报文的转发。策略路由用 MATCH 和 SET 语句实现路径的选择。策略路由设置在接收报文接口上而不是发送接口上。

任务实现

步骤 1:配置路由器 R1 的基础网络环境。

```
Router_config#hostname R1
R1_config#interface fastEthernet 0/0
R1_config_f0/0#ip address 1.1.3.1 255.255.255.0
R1_config_f0/0#exit
R1_config#interface fastEthernet 0/3
```

```
R1_config_f0/3#ip address 1.1.2.1 255.255.255.0
R1_config_f0/3#exit
R1_config#interface loopback 0
R1_config_l0#ip address 1.1.1.1 255.255.255.0
R1_config_l0#exit
R1_config#
```

步骤 2：配置路由器 R2 的基础网络环境。

```
Router_config#hostname R2
R2_config#interface fastEthernet 0/0
R2_config_f0/0#ip address 1.1.3.2 255.255.255.0
R2_config_f0/0#exit
R2_config#interface serial 0/3
R2_config_s0/3#physical-layer speed 64000
R2_config_s0/3#ip address 1.1.5.1 255.255.255.0
R2_config_s0/3#exit
R2_config#
```

步骤 3：配置路由器 R3 的基础网络环境。

```
Router_config#hostname R3
R3_config#interface fastEthernet 0/0
R3_config_f0/0#ip address 1.1.2.2 255.255.255.0
R3_config_f0/0#exit
R3_config#interface fastEthernet 0/3
R3_config_f0/3#ip address 1.1.4.1 255.255.255.0
R3_config_f0/3#exit
R3_config#interface serial 0/2
R3_config_s0/2#ip address 1.1.5.2 255.255.255.0
R3_config#
```

步骤 4：测试 R2 链路的连通性。

```
R2#ping 1.1.3.1
PING 1.1.3.1 (1.1.3.1): 56 data bytes
!!!!!
--- 1.1.3.1 ping statistics ---
5 packets transmitted, 5 packets received, 0% packet loss
round-trip min/avg/max = 0/0/0 ms
R2#ping 1.1.5.2
PING 1.1.5.2 (1.1.5.2): 56 data bytes
!!!!!
--- 1.1.5.2 ping statistics ---
5 packets transmitted, 5 packets received, 0% packet loss
round-trip min/avg/max = 0/0/0 ms
R2#
```

步骤 5：测试 R3 链路的连通性。

```
R3#ping 1.1.2.1
PING 1.1.2.1 (1.1.2.1): 56 data bytes
```

!!!!!
--- 1.1.2.1 ping statistics ---
5 packets transmitted, 5 packets received, 0% packet loss
round-trip min/avg/max = 0/0/0 ms
R3#

这表示单条链路都可以连通。

步骤 6：配置 R1 路由环境，使用 OSPF 单区域配置。

R1_config#router ospf 1
R1_config_ospf_1#network 1.1.3.0 255.255.255.0 area 0
R1_config_ospf_1#network 1.1.2.0 255.255.255.0 area 0
R1_config_ospf_1#redistribute connect
R1_config_ospf_1#exit
R1_config#

步骤 7：配置 R2 路由环境，使用 OSPF 单区域配置。

R2_config#router ospf 1
R2_config_ospf_1#network 1.1.3.0 255.255.255.0 area 0
R2_config_ospf_1#network 1.1.5.0 255.255.255.0 area 0
R2_config_ospf_1#redistribute connect
R2_config_ospf_1#exit
R2_config#

步骤 8：配置 R3 路由环境，使用 OSPF 单区域配置。

R3_config#router ospf 1
R3_config_ospf_1#network 1.1.2.0 255.255.255.0 area 0
R3_config_ospf_1#network 1.1.5.0 255.255.255.0 area 0
R3_config_ospf_1#exit
R3_config#router ospf 1
R3_config_ospf_1#redistribute connect
R3_config_ospf_1#exit
R3_config#

步骤 9：查看 R1 的路由表。

R1#sh ip route
Codes: C - connected, S - static, R - RIP, B - BGP, BC - BGP connected
       D - DEIGRP, DEX - external DEIGRP, O - OSPF, OIA - OSPF inter area
       ON1 - OSPF NSSA external type 1, ON2 - OSPF NSSA external type 2
       OE1 - OSPF external type 1, OE2 - OSPF external type 2
       DHCP - DHCP type
VRF ID: 0
C   1.1.1.0/24   is directly connected, Loopback0
C   1.1.2.0/24   is directly connected, FastEthernet0/3
C   1.1.3.0/24   is directly connected, FastEthernet0/0
O E2   1.1.4.0/24   [150,100] via 1.1.2.2(on FastEthernet0/3)
O   1.1.5.0/24   [110,1601] via 1.1.2.2(on FastEthernet0/3)
R1#

步骤 10：查看 R2 的路由表。

```
R2#sh ip route
Codes: C - connected, S - static, R - RIP, B - BGP, BC - BGP connected
       D - DEIGRP, DEX - external DEIGRP, O - OSPF, OIA - OSPF inter area
       ON1 - OSPF NSSA external type 1, ON2 - OSPF NSSA external type 2
       OE1 - OSPF external type 1, OE2 - OSPF external type 2
       DHCP - DHCP type
VRF ID: 0
O E2   1.1.1.0/24        [150,100] via 1.1.3.1(on FastEthernet0/0)
O      1.1.2.0/24        [110,2] via 1.1.3.1(on FastEthernet0/0)
C      1.1.3.0/24        is directly connected, FastEthernet0/0
O E2   1.1.4.0/24        [150,100] via 1.1.3.1(on FastEthernet0/0)
C      1.1.5.0/24        is directly connected, Serial0/3
R2#
```

步骤 11：查看 R3 的路由表。

```
R3#sh ip route
Codes: C - connected, S - static, R - RIP, B - BGP, BC - BGP connected
       D - DEIGRP, DEX - external DEIGRP, O - OSPF, OIA - OSPF inter area
       ON1 - OSPF NSSA external type 1, ON2 - OSPF NSSA external type 2
       OE1 - OSPF external type 1, OE2 - OSPF external type 2
       DHCP - DHCP type
VRF ID: 0
O E2   1.1.1.0/24        [150,100] via 1.1.2.1(on FastEthernet0/0)
C      1.1.2.0/24        is directly connected, FastEthernet0/0
O      1.1.3.0/24        [110,2] via 1.1.2.1(on FastEthernet0/0)
C      1.1.4.0/24        is directly connected, FastEthernet0/3
C      1.1.5.0/24        is directly connected, Serial0/2
R3#
```

步骤 12：在 R3 中使用策略路由。使来自 1.1.4.10 的源地址去往外网的路由从 1.1.2.1 通过，而来自 1.1.4.20 的源地址的数据从 1.1.5.1 的路径通过。

```
R3_config#ip access-list standard for_10
R3_config_std_nacl#permit 1.1.4.10
R3_config_std_nacl#exit
R3_config#ip access-list standard for_20
R3_config_std_nacl#permit 1.1.4.20
R3_config_std_nacl#exit
R3_config#route-map source_pbr 10 permit
R3_config_route_map#match ip address for_10
R3_config_route_map#set ip next-hop 1.1.2.1
R3_config_route_map#exit
R3_config#route-map source_pbr 20 permit
R3_config_route_map#match ip address for_20
R3_config_route_map#set ip next-hop 1.1.5.1
R3_config_route_map#exit
R3_config#interface fastEthernet 0/3
```

```
R3_config_f0/3#ip policy route-map source_pbr
R3_config_f0/3#
```

此时已经更改了 R3 的路由策略，终端测试的结果如下。

```
------------------------1.1.4.10------------------------------
C:\Documents and Settings\Administrator>ipconfig
Windows IP Configuration
Ethernet adapter 本地连接:
        Connection-specific DNS Suffix   . :
        IP Address. . . . . . . . . . . : 1.1.4.10
        Subnet Mask . . . . . . . . . . : 255.255.255.0
        Default Gateway . . . . . . . . : 1.1.4.1
C:\Documents and Settings\Administrator>tracert 1.1.1.1
Tracing route to 1.1.1.1 over a maximum of 30 hops
    1     <1 ms<1 ms<1 ms    1.1.4.1
    2      1 ms<1 ms<1 ms    1.1.1.1
Trace complete.
C:\Documents and Settings\Administrator>
C:\>ipconfig
Windows IP Configuration
------------------------1.1.4.20------------------------------
Ethernet adapter 本地连接:
        Connection-specific DNS Suffix   . :
        IP Address. . . . . . . . . . . : 1.1.4.20
        Subnet Mask . . . . . . . . . . : 255.255.255.0
        Default Gateway . . . . . . . . : 1.1.4.1
C:\>tracert 1.1.1.1
Tracing route to 1.1.1.1 over a maximum of 30 hops
    1     <1 ms<1 ms<1 ms    1.1.4.1
    2     16 ms    15 ms     15 ms    1.1.5.1
    3     15 ms    14 ms     15 ms    1.1.1.1
Trace complete.
```

此时可以看出，不同源的路由已经发生了改变。

任务四 基于应用的策略路由

需求分析

由于公司员工人数很多，网络出口的链路带宽和花销也不同，为了有效地提高链路的使用效率，使员工都可以满意的办公，网络管理员想尝试新的办法。

方案设计

当网络的出口链路带宽和花销不同时，将关键业务的流量分配给带宽大的链路负载，将不重要且不紧急的流量分配给带宽小的链路处理，可以有效地提高链路的使用效率。

所需设备如图 4-4-1 所示。
（1）DCR-2626 路由器 3 台。
（2）PC 2 台。
（3）网线若干。

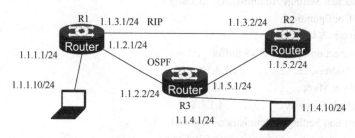

图 4-4-1　基于应用的策略路由

任务要求：配置基础网络环境，在网络中使用静态路由，在 R3 中不使用任何静态路由，只在 R1 和 R2 中定义静态路由，R1 定义时使用浮动路由的配置，即到 1.1.4.0 网络的路由存在两个，但其度量值不同。在 R3 中启动策略路由，定义 TCP 数据下一跳为 1.1.2.1，UDP 数据下一跳为 1.1.5.2。从 1.1.1.10 中启动一个 FTP 服务，从 1.1.4.10 发起一次 FTP 请求，在中间过程中将 1.1.5.1 端口停用，查看传输是否有影响，再启用此接口，停用 1.1.2.2 接口，结果如何？本任务由于涉及应用类型，因此，访问控制列表必须使用扩展列表，标准访问控制列表是无法实现的。由于 R3 路由器的静态路由针对 1.1.1.0 网络是指向 1.1.2.1 的，因此本任务应将策略路由的指向改为 1.1.5.1，目的是方便观测试验效果，但在实际应用中应尽量不使用次优的路由定义策略。

♂ 知识准备

应用策略路由时，必须指定策略路由使用的路由图，并且需要创建路由图。一个路由图由很多条策略组成，每个策略都定义了一个或多个的匹配规则和对应操作。一个接口应用策略路由后，将对该接口接收到的所有包进行检查，不符合路由图任意策略的数据包将按照通常的路由转发进行处理，符合路由图中某个策略的数据包就按照该策略中定义的操作进行处理。

♂ 任务实现

步骤 1：配置基础网络，此处略。
步骤 2：配置 R1 的静态路由。

```
R1_config#ip route 1.1.4.0 255.255.255.0 1.1.2.2
R1_config#ip route 1.1.5.0 255.255.255.0 1.1.3.2
R1_config#ip route 1.1.5.0 255.255.255.0 1.1.2.2
R1_config#
```

步骤 3：配置 R2 的静态路由。

```
R2_config#ip route 1.1.2.0 255.255.255.0 1.1.3.1
R2_config#ip route 1.1.1.0 255.255.255.0 1.1.3.1
R2_config#ip route 1.1.4.0 255.255.255.0 1.1.5.2
R2_config#ip route 1.1.2.0 255.255.255.0 1.1.5.2
```

步骤 4：配置 R3 的静态路由。

项目四 路由重分布与策略路由

```
R3_config#ip route 1.1.3.0 255.255.255.0 1.1.5.1
R3_config#ip route 1.1.3.0 255.255.255.0 1.1.2.1
R3_config#ip route 1.1.1.0 255.255.255.0 1.1.2.1
```

步骤 5：查看 R1 的路由表。

```
R1#sh ip route
Codes: C - connected, S - static, R - RIP, B - BGP, BC - BGP connected
       D - DEIGRP, DEX - external DEIGRP, O - OSPF, OIA - OSPF inter area
       ON1 - OSPF NSSA external type 1, ON2 - OSPF NSSA external type 2
       OE1 - OSPF external type 1, OE2 - OSPF external type 2
       DHCP - DHCP type
VRF ID: 0
C    1.1.1.0/24        is directly connected, Loopback0
C    1.1.2.0/24        is directly connected, FastEthernet0/3
C    1.1.3.0/24        is directly connected, FastEthernet0/0
S    1.1.4.0/24        [1,0] via 1.1.2.2(on FastEthernet0/3)
S    1.1.5.0/24        [1,0] via 1.1.2.2(on FastEthernet0/3)
                       [1,0] via 1.1.3.2(on FastEthernet0/0)
R1#
```

步骤 6：查看 R2 的路由表。

```
R2#sh ip route
Codes: C - connected, S - static, R - RIP, B - BGP, BC - BGP connected
       D - DEIGRP, DEX - external DEIGRP, O - OSPF, OIA - OSPF inter area
       ON1 - OSPF NSSA external type 1, ON2 - OSPF NSSA external type 2
       OE1 - OSPF external type 1, OE2 - OSPF external type 2
       DHCP - DHCP type
VRF ID: 0
S    1.1.1.0/24        [1,0] via 1.1.3.1(on FastEthernet0/0)
S    1.1.2.0/24        [1,0] via 1.1.3.1(on FastEthernet0/0)
                       [1,0] via 1.1.5.2(on Serial0/3)
C    1.1.3.0/24        is directly connected, FastEthernet0/0
S    1.1.4.0/24        [1,0] via 1.1.5.2(on Serial0/3)
C    1.1.5.0/24        is directly connected, Serial0/3
R2#
```

步骤 7：查看 R3 的路由表。

```
R3#sh ip rouie
Codes: C - connected, S - static, R – RIP, B - BGP, BC - BGP connected
       D - DEIGRP, DEX - external DEIGRP, O - OSPF, OIA - OSPF inter area
       ON1 - OSPF NSSA external type 1, ON2 - OSPF NSSA external type 2
       OE1 - OSPF external type 1, OE2 - OSPF external type 2
       DHCP - DHCP type
VRF ID: 0
S    1.1.1.0/24        [1,0] via 1.1.2.1(on FastEthernet0/0)
C    1.1.2.0/24        is directly connected, FastEthernet0/0
S    1.1.3.0/24        [1,0] via 1.1.2.1(on FastEthernet0/0)
```

		[1,0] via 1.1.5.1(on Serial0/2)
C	1.1.4.0/24	is directly connected, FastEthernet0/3
C	1.1.5.0/24	is directly connected, Serial0/2

R3#

步骤 8：配置 R3 的策略路由，定义 ICMP 数据下一跳为 1.1.5.1，UDP 数据下一跳为 1.1.2.1。

R3_config#ip access-list extended for_icmp
R3_config_ext_nacl#permit icmp any any
R3_config_ext_nacl#exit
R3_config#route-map app_pbr 10 permit
R3_config_route_map#match ip address for_icmp
R3_config_route_map#set ip next-hop 1.1.5.1
R3_config_route_map#exit
R3_config#route-map app_pbr 20 permit
R3_config_route_map# match ip address for_udp
R3_config_route_map#set ip next-hop 1.1.2.1
R3_config_route_map#exit
R3_config#R3_config#interface fastEthernet 0/3
R3_config_f0/3#ip policy route-map app_pbr
R3_config_f0/3#

步骤 9：查看当前配置情况。

R3_config#sh ip policy
 Interface   Route-map
 FastEthernet0/3   app_pbr
R3_config#

步骤 10：测试结果。从 PC 上测试策略路由的配置是否生效了，使用如下过程：使用 ping 1.1.1.1 命令来测试整个过程，在策略路由生效的情况下，返回连通状态。

C:\Documents and Settings\Administrator>ping 1.1.1.1
Pinging 1.1.1.1 with 32 bytes of data:
Reply from 1.1.1.1: bytes=32 time=12ms TTL=254
Reply from 1.1.1.1: bytes=32 time=15ms TTL=254
Reply from 1.1.1.1: bytes=32 time=13ms TTL=254
Reply from 1.1.1.1: bytes=32 time=11ms TTL=254
Ping statistics for 1.1.1.1:
 Packets: Sent = 4, Received = 4, Lost = 0 (0% loss),
Approximate round trip times in milli-seconds:
 Minimum = 11ms, Maximum = 15ms, Average = 12ms
C:\Documents and Settings\Administrator>

此时将策略路由指向 1.1.5.1 的链路断开（可使用 shutdown 命令），由于策略路由不存在，而 R3 的静态路由还在生效，因此依然是连通状态。但通过仔细观察，可发现其返回时间是不同的。

C:\Documents and Settings\Administrator>ping 1.1.1.1
Pinging 1.1.1.1 with 32 bytes of data:
Reply from 1.1.1.1: bytes=32 time=1ms TTL=254

```
Reply from 1.1.1.1: bytes=32 time<1ms TTL=254
Reply from 1.1.1.1: bytes=32 time<1ms TTL=254
Reply from 1.1.1.1: bytes=32 time<1ms TTL=254
Ping statistics for 1.1.1.1:
    Packets: Sent = 4, Received = 4, Lost = 0 (0% loss),
Approximate round trip times in milli-seconds:
    Minimum = 0ms, Maximum = 1ms, Average = 0ms
C:\Documents and Settings\Administrator>
```

此时，将静态路由删除，命令如下：

```
R3_config#no ip route 1.1.1.0 255.255.255.0 1.1.2.1
```

再次测试连通性，结果如下：

```
C:\Documents and Settings\Administrator>ping 1.1.1.1
Pinging 1.1.1.1 with 32 bytes of data:
Reply from 1.1.4.1: Destination host unreachable.
Reply from 1.1.4.1: Destination host unreachable.
Reply from 1.1.4.1: Destination host unreachable.
Reply from 1.1.4.1: Destination host unreachable.
Ping statistics for 1.1.1.1:
    Packets: Sent = 4, Received = 4, Lost = 0 (0% loss)
Approximate round trip times in milli-seconds:
    Minimum = 0ms, Maximum = 0ms, Average = 0ms
C:\Documents and Settings\Administrator>
```

此时发现既没有策略路由，又没有静态路由，已经无法连通了。此时将策略路由所使用的链路启动，再次测试。命令如下：

```
R3_config#int s 0/2
R3_config_s0/2#no shut
R3_config_s0/2#Jan   1 03:42:17 Line on Interface Serial0/2, changed to up
Jan   1 03:42:27 Line protocol on Interface Serial0/2, change state to up
```

测试结果如下：

```
C:\Documents and Settings\Administrator>ping 1.1.1.1
Pinging 1.1.1.1 with 32 bytes of data:
Reply from 1.1.1.1: bytes=32 time=12ms TTL=254
Reply from 1.1.1.1: bytes=32 time=11ms TTL=254
Reply from 1.1.1.1: bytes=32 time=11ms TTL=254
Reply from 1.1.1.1: bytes=32 time=11ms TTL=254
Ping statistics for 1.1.1.1:
    Packets: Sent = 4, Received = 4, Lost = 0 (0% loss),
Approximate round trip times in milli-seconds:
    Minimum = 11ms, Maximum = 12ms, Average = 11ms
C:\Documents and Settings\Administrator>
```

注意，此时的返回时间与第一次测试时是一致的。

认证考核

实训题

背景与需求：某公司总部位于北京，在上海、广州有分公司。总部与分公司通过路由器相连，运行的是 OSPF 路由协议，总部内部采用的是 RIP 路由协议。网络结构如图 4-4-2 所示，图中的 SH 代表上海，GZ 代表广州，BJ 代表北京。作为网络管理员要完成总公司到分公司的互连，在此采用了重分布技术。注意，GZ 路由器上的环回地址与 BJ 路由器采用 RIP 技术互连。

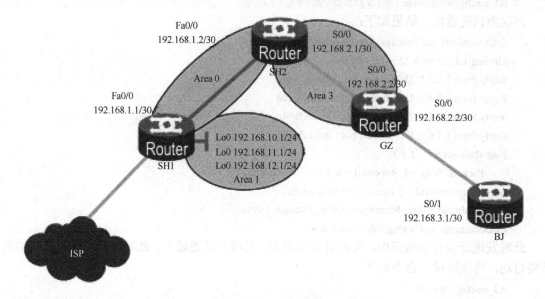

图 4-4-2 网络结构

（1）配置多区域 OSPF。
（2）配置 RIP。
（3）配置 RIP 和 OSPF 的重分布。
完成标准：网络正确连通；在 SH1 上能够使用 show ip route 命令查看到 BJ 的 RIP 的路由。

项目五 访问控制列表

♂ 教学背景

由于网络的迅猛发展,网络安全已成为网络发展的一个重要议题。企业网络的安全问题也日益突出,忽视了网络设备的安全设置,就如同整个网络敞开了大门一样,不过,通过针对性的设置可以使它变成一堵安全的"墙"。

任务一 标准 ACL

♂ 需求分析

某公司组建了公司内部网络,其中有财务部、技术部、市场部等部门。为了保证公司财务安全,公司经理让网络管理员禁止技术部门的员工访问财务部门的主机。

♂ 方案设计

ACL(Access Control List,访问控制列表)是交换机实现的一种数据包过滤机制,通过允许或拒绝特定的数据包进出网络,交换机可以对网络访问进行控制,有效保证网络的安全运行。用户可以基于报文中的特定信息制定一组规则,每条规则都描述了对匹配一定信息的数据包所采取的动作:允许通过(Permit)或拒绝通过(Deny)。用户可以把这些规则应用到特定交换机端口的入口或出口方向,这样特定端口上特定方向的数据流就必须依照指定的 ACL 规则进出交换机。通过 ACL,可以限制某个 IP 地址的 PC 或者某些网段的 PC 的上网活动,有效实现网络管理。

所需设备如图 5-1-1 所示。
(1) DCRS-5650 交换机 2 台。
(2) PC 2 台。
(3) Console 线 1 或 2 条。
(4) 直通网线若干。

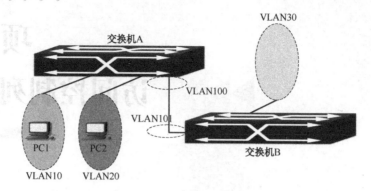

图 5-1-1 标准 ACL 拓扑

（1）在交换机 A 和交换机 B 上分别划分基于端口的 VLAN，见表 5-1-1。

表 5-1-1 交换机 VLAN 的划分

交换机	VLAN	端口成员
交换机 A	10	1~8
	20	9~16
	100	24
交换机 B	30	1~8
	101	24

（2）交换机 A 和交换机 B 通过 24 口级联。

（3）配置交换机 A 和交换机 B 各 VLAN 虚拟接口的 IP 地址，见表 5-1-2。

表 5-1-2 各 VLAN 虚拟接口的 IP 地址

VLAN	VLAN10	VLAN20	VLAN30	VLAN100	VLAN101
IP 地址	192.168.10.1	192.168.20.1	192.168.30.1	192.168.100.1	192.168.100.2

（4）PC1 和 PC2 的网络设置见表 5-1-3。

表 5-1-3 PC1 和 PC2 的网络设置

设备	IP 地址	网关	子网掩码
PC1	192.168.10.101	192.168.10.1	255.255.255.0
PC2	192.168.20.101	192.168.20.1	255.255.255.0

（5）验证：PC1 和 PC2 都通过交换机 A 连接到交换机 B，若不配置 ACL，两台 PC 都可以连通 VLAN 30；配置 ACL 后，PC1 和 PC2 无法连通 VLAN 30，更改了 IP 地址后即可以连通。

若测试结果和理论相符，则本任务完成。

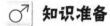 知识准备

IP 访问控制列表通过对数据流进行检查和过滤限制网络中通信数据的类型，限制网络用户及用户所访问的设备。ACL 由一系列有序的 ACE 组成，每一个 ACE 都定义了匹配条件及行为。

标准 ACL 只能针对源 IP 地址制定匹配条件，对于符合匹配条件的数据包，ACE 执行所规定的行为：允许或拒绝。

在设备的入站方向或者出站方向上可以应用 ACL。如果在设备的入站方向上应用 ACL，则设备在端口上收到数据包后，先进行 ACL 规定的检查。检查从 ACL 的第一个 ACE 开始，将 ACE 规定的条件和数据包内容进行比较匹配。如果第一个 ACE 没有匹配成功，则匹配下一个 ACE，以此类推。一旦匹配成功，则执行该 ACE 规定的行为。如果整个 ACL 中所有 ACE 都没有匹配成功，则执行设备定义的默认行为。被 ACL 放行的数据包会进一步执行设备的其他策略，如路由转发。

如果路由器某接口出站方向应用了 ACL，则路由器会先进行路由转发决策，发送到该接口的数据包应用 ACL 检查，检查过程与入站方向一致。

定义 ACL 应当遵循以下规则。

（1）设备接口的一个方向只能应用一个 ACL。
（2）ACL 匹配自顶向下，逐条匹配。
（3）一旦某一个 ACE 匹配成功，就立即执行该 ACE 的行为，停止匹配。
（4）如果所有 ACE 都没有匹配成功，则执行设备定义的默认行为。
（5）一般情况下，在什么设备上、什么接口上、什么方向上应用 ACL，必须遵循以下约定。
  ① 标准 ACL 一般应用在离数据流的目的地尽可能近的地方。
  ② 扩展 ACL 一般应用在离数据流的源地址尽可能近的地方。

尽管大部分厂商推出了更高级的 ACL，但是绝大部分的网络管理员只使用两种 ACL：标准 ACL 和扩展 ACL。尽管功能比较简单，但标准 ACL 在限制 Telnet 访问路由器、限制通过 HTTP 访问设备、路由过滤更新方面，仍然有着较大的应用。

标准 ACL 的编号为 1～99、1300～1999。

在全局配置模式下，配置编号的标准 ACL 语法如下。

access-list<num> {deny | permit} {{<sIpAddr><sMask>} | any-source | {host-source <sIpAddr>}}

该命令用于创建一个数字标准 IP 访问列表。如果该列表已经存在，则在该 ACL 中增加一条 ACE。使用"no access-list <num>"可以删除一个 ACL。

一般情况下，配置 ACL 应遵循以下步骤。

（1）启用设备包过滤功能并配置默认行为。
（2）定义 ACL 规则。
（3）绑定 ACL 到设备接口的某一方向上。

♂ 任务实现

步骤 1：将交换机全部恢复为出厂设置。
步骤 2：交换机 A 的 VLAN 配置。

```
DCRS-5650-A(Config)#vlan 10
DCRS-5650-A(Config-Vlan10)#switchport interface ethernet 0/0/1-8
Set the port Ethernet0/0/1 access vlan 10 successfully
Set the port Ethernet0/0/2 access vlan 10 successfully
Set the port Ethernet0/0/3 access vlan 10 successfully
```

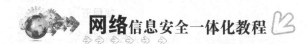

```
Set the port Ethernet0/0/4 access vlan 10 successfully
Set the port Ethernet0/0/5 access vlan 10 successfully
Set the port Ethernet0/0/6 access vlan 10 successfully
Set the port Ethernet0/0/7 access vlan 10 successfully
Set the port Ethernet0/0/8 access vlan 10 successfully
DCRS-5650-A(Config-Vlan10)#exit
DCRS-5650-A(Config)#vlan 20
DCRS-5650-A(Config-Vlan20)#switchport interface ethernet 0/0/9-16
Set the port Ethernet0/0/9 access vlan 20 successfully
Set the port Ethernet0/0/10 access vlan 20 successfully
Set the port Ethernet0/0/11 access vlan 20 successfully
Set the port Ethernet0/0/12 access vlan 20 successfully
Set the port Ethernet0/0/13 access vlan 20 successfully
Set the port Ethernet0/0/14 access vlan 20 successfully
Set the port Ethernet0/0/15 access vlan 20 successfully
Set the port Ethernet0/0/16 access vlan 20 successfully
DCRS-5650-A(Config-Vlan20)#exit
DCRS-5650-A(Config)#vlan 100
DCRS-5650-A(Config-Vlan100)#switchport interface ethernet 0/0/24
Set the port Ethernet0/0/24 access vlan 100 successfully
DCRS-5650-A(Config-Vlan100)#exit
DCRS-5650-A(Config)#
```

步骤 3：交换机 A 的 VLAN 配置的验证。

```
DCRS-5650-A#show vlan
VLAN Name           Type      Media    Ports
-----------------------------------------------------------
1    default        Static    ENET     Ethernet0/0/17    Ethernet0/0/18
                                       Ethernet0/0/19    Ethernet0/0/20
                                       Ethernet0/0/21    Ethernet0/0/22
                                       Ethernet0/0/23    Ethernet0/0/25
                                       Ethernet0/0/26    Ethernet0/0/27
                                       Ethernet0/0/28
10   VLAN0010       Static    ENET     Ethernet0/0/1     Ethernet0/0/2
                                       Ethernet0/0/3     Ethernet0/0/4
                                       Ethernet0/0/5     Ethernet0/0/6
                                       Ethernet0/0/7     Ethernet0/0/8
20   VLAN0020       Static    ENET     Ethernet0/0/9     Ethernet0/0/10
                                       Ethernet0/0/11    Ethernet0/0/12
                                       Ethernet0/0/13    Ethernet0/0/14
                                       Ethernet0/0/15    Ethernet0/0/16
100  VLAN0100       Static    ENET     Ethernet0/0/24
DCRS-5650-A#
```

步骤 4：交换机 B 的 VLAN 配置。

```
DCRS-5650-B(Config)#vlan 30
```

```
DCRS-5650-B(Config-Vlan30)#switchport interface ethernet 0/0/1-8
Set the port Ethernet0/0/1 access vlan 30 successfully
Set the port Ethernet0/0/2 access vlan 30 successfully
Set the port Ethernet0/0/3 access vlan 30 successfully
Set the port Ethernet0/0/4 access vlan 30 successfully
Set the port Ethernet0/0/5 access vlan 30 successfully
Set the port Ethernet0/0/6 access vlan 30 successfully
Set the port Ethernet0/0/7 access vlan 30 successfully
Set the port Ethernet0/0/8 access vlan 30 successfully
DCRS-5650-B(Config-Vlan30)#exit
DCRS-5650-B(Config)#vlan 40
DCRS-5650-B(Config-Vlan40)#switchport interface ethernet 0/0/9-16
Set the port Ethernet0/0/9 access vlan 40 successfully
Set the port Ethernet0/0/10 access vlan 40 successfully
Set the port Ethernet0/0/11 access vlan 40 successfully
Set the port Ethernet0/0/12 access vlan 40 successfully
Set the port Ethernet0/0/13 access vlan 40 successfully
Set the port Ethernet0/0/14 access vlan 40 successfully
Set the port Ethernet0/0/15 access vlan 40 successfully
Set the port Ethernet0/0/16 access vlan 40 successfully
DCRS-5650-B(Config-Vlan40)#exit
DCRS-5650-B(Config)#vlan 101
DCRS-5650-B(Config-Vlan101)#switchport interface ethernet 0/0/24
Set the port Ethernet0/0/24 access vlan 101 successfully
DCRS-5650-B(Config-Vlan101)#exit
DCRS-5650-B(Config)#
```

步骤 5：交换机 B 的 VLAN 配置的验证。

```
DCRS-5650-B#show vlan
```

VLAN	Name	Type	Media	Ports	
1	default	Static	ENET	Ethernet0/0/17	Ethernet0/0/18
				Ethernet0/0/19	Ethernet0/0/20
				Ethernet0/0/21	Ethernet0/0/22
				Ethernet0/0/23	Ethernet0/0/25
				Ethernet0/0/26	Ethernet0/0/27
				Ethernet0/0/28	
10	VLAN0010	Static	ENET	Ethernet0/0/1	Ethernet0/0/2
				Ethernet0/0/3	Ethernet0/0/4
				Ethernet0/0/5	Ethernet0/0/6
				Ethernet0/0/7	Ethernet0/0/8
20	VLAN0020	Static	ENET	Ethernet0/0/9	Ethernet0/0/10
				Ethernet0/0/11	Ethernet0/0/12
				Ethernet0/0/13	Ethernet0/0/14
				Ethernet0/0/15	Ethernet0/0/16

```
100     VLAN0100        Static      ENET        Ethernet0/0/24
DCRS-5650-B#
```

步骤 6：配置交换机 A 的 VLAN 虚拟接口的 IP 地址。

```
DCRS-5650-A(Config)#int vlan 10
DCRS-5650-A(Config-If-Vlan10)#ip address 192.168.10.1 255.255.255.0
DCRS-5650-A(Config-If-Vlan10)#no shut
DCRS-5650-A(Config-If-Vlan10)#exit
DCRS-5650-A(Config)#int vlan 20
DCRS-5650-A(Config-If-Vlan20)#ip address 192.168.20.1 255.255.255.0
DCRS-5650-A(Config-If-Vlan20)#no shut
DCRS-5650-A(Config-If-Vlan20)#exit
DCRS-5650-A(Config)#int vlan 100
DCRS-5650-A(Config-If-Vlan100)#ip address 192.168.100.1 255.255.255.0
DCRS-5650-A(Config-If-Vlan100)#no shut
DCRS-5650-A(Config-If-Vlan100)#exit
DCRS-5650-A(Config)#
```

步骤 7：配置交换机 B 的 VLAN 虚拟接口的 IP 地址。

```
DCRS-5650-B(Config)#int vlan 30
DCRS-5650-B(Config-If-Vlan30)#ip address 192.168.30.1 255.255.255.0
DCRS-5650-B(Config-If-Vlan30)#no shut
DCRS-5650-B(Config-If-Vlan30)#exit
DCRS-5650-B(Config)#int vlan 101
DCRS-5650-B(Config-If-Vlan101)#ip address 192.168.100.2 255.255.255.0
DCRS-5650-B(Config-If-Vlan101)#exit
DCRS-5650-B(Config)#
```

步骤 8：配置静态路由。

```
DCRS-5650-A(Config)#ip route 0.0.0.0 0.0.0.0 192.168.100.2
DCRS-5650-B(Config)#ip route 0.0.0.0 0.0.0.0 192.168.100.1
```

步骤 9：查看交换机 A 的路由表。

```
DCRS-5650-A#show ip route
Codes: K - kernel, C - connected, S - static, R - RIP, B - BGP
       O - OSPF, IA - OSPF inter area
       N1 - OSPF NSSA external type 1, N2 - OSPF NSSA external type 2
       E1 - OSPF external type 1, E2 - OSPF external type 2
       i - IS-IS, L1 - IS-IS level-1, L2 - IS-IS level-2, ia - IS-IS inter area
       * - candidate default
Gateway of last resort is 192.168.100.2 to network 0.0.0.0
S*        0.0.0.0/0 [1/0] via 192.168.100.2, Vlan100
C         127.0.0.0/8 is directly connected, Loopback
C         192.168.10.0/24 is directly connected, Vlan10
C         192.168.20.0/24 is directly connected, Vlan10
C         192.168.100.0/24 is directly connected, Vlan100
```

步骤 10：在 VLAN 30 端口上配置端口的环回测试功能，保证 VLAN 30 可以连通。

```
DCRS-5650-B(Config)# interface ethernet 0/0/1         //任意一个 VLAN 30 内的接口均可
```

```
DCRS-5650-B(Config-If-Ethernet0/0/1)#loopback
DCRS-5650-B(Config-If-Ethernet0/0/1)#no shut
DCRS-5650-B(Config-If-Ethernet0/0/1)#exit
```

步骤 11：不配置 ACL 验证结果。验证 PC1 和 PC2 之间是否可以连通 VLAN 30 的虚接口的 IP 地址。

步骤 12：配置访问控制列表。

```
DCRS-5650-A(Config)#ip access-list standard test
DCRS-5650-A(Config-Std-Nacl-test)#deny 192.168.10.101 0.0.0.255
DCRS-5650-A(Config-Std-Nacl-test)#deny host-source192.168.20.101
DCRS-5650-A(Config-Std-Nacl-test)#exit
DCRS-5650-A(Config)#
```

步骤 13：查看已经配置的 ACL。

```
DCRS-5650-A#show access-lists
ip access-list standard test(used 1 time(s))
deny 192.168.10.101 0.0.0.255
deny host-source 192.168.20.101
```

步骤 14：配置访问控制列表功能启用，默认动作为全部启用。

```
DCRS-5650-A(Config)#firewall enable
DCRS-5650-A(Config)#firewall default permit
DCRS-5650-A(Config)#
```

验证配置：

```
DCRS-5650-A#show firewall
Fire wall is enabled.
Firewall default rule is to permit any ip packet.
DCRS-5650-A#
```

步骤 15：绑定 ACL 到各端口上。

```
DCRS-5650-A(Config)#interface ethernet0/0/1
DCRS-5650-A(Config-Ethernet0/0/1)#ip access-group 11 in
DCRS-5650-A(Config-Ethernet0/0/1)#exit
DCRS-5650-A(Config)#interface ethernet0/0/9
DCRS-5650-A(Config-Ethernet0/0/9)#ip access-group 11 in
DCRS-5650-A(Config-Ethernet0/0/9)#exit
```

步骤 16：查看已经配置的 ACL。

```
DCRS-5650-A#show access-group
interface name:Ethernet0/0/9
    IP Ingress access-list used is 11, traffic-statistics Disable.
interface name:Ethernet0/0/1
    IP Ingress access-list used is 11, traffic-statistics Disable.
```

步骤 17：验证结果，见表 5-1-4。

表 5-1-4　验证结果

PC	端口	ping	结果	原因
PC1：192.168.10.101	0/0/1	192.168.30.1	不通	
PC1：192.168.10.12	0/0/1	192.168.30.1	通	
PC2：192.168.20.101	0/0/9	192.168.30.1	不通	
PC2：192.168.20.12	0/0/9	192.168.30.1	通	

任务二　扩展 ACL

需求分析

某公司为了提升工作效率，购买了专门的服务器，用于搭建公司的办公自动化系统。基于安全上的考虑，公司希望禁止销售部门的员工对服务器的 FTP 服务，销售部门的员工只能访问公司服务器的其他服务。

方案设计

标准 ACL 只能限制源 IP 地址，而扩展 ACL 的限制权限就很广泛了，包括源 IP、目的 IP、服务类型等。

所需设备如图 5-2-1 所示。
（1）DCRS-5650 交换机 2 台。
（2）PC 2 台。
（3）Console 线 1 或 2 条。
（4）直通网线若干。

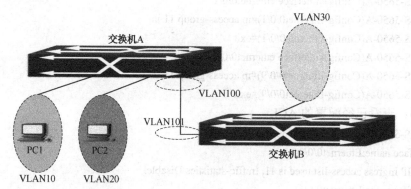

图 5-2-1　扩展 ACL 拓扑

目的：禁止 PC2 连通 VLAN30。
（1）在交换机 A 和交换机 B 上分别划分基于端口的 VLAN，见表 5-2-1。

表 5-2-1　交换机 VLAN 的划分

交换机	VLAN	端口成员
交换机 A	10	1～8
	20	9～16
	100	24
交换机 B	30	1～8
	101	24

（2）交换机 A 和交换机 B 通过 24 口级联。
（3）配置交换机 A 和交换机 B 各 VLAN 虚拟接口的 IP 地址，见表 5-2-2。

表 5-2-2　各 VLAN 的 IP 地址

VLAN	VLAN10	VLAN20	VLAN30	VLAN100	VLAN101
IP 地址	192.168.10.1	192.168.20.1	192.168.30.1	192.168.100.1	192.168.100.2

（4）PC1 和 PC2 的网络设置见表 5-2-3。

表 5-2-3　PC1 和 PC2 的网络设置

设备	IP 地址	网关	子网掩码
PC1	192.168.10.101	192.168.10.1	255.255.255.0
PC2	192.168.20.101	192.168.20.1	255.255.255.0

验证：
（1）配置 ACL 之前，PC1 和 PC2 都可以连通 VLAN30。
（2）配置 ACL 后，PC1 可以连通 VLAN30，而 PC2 不可以连通 VLAN30。
若测试结果和理论相符，则本任务完成。

知识准备

与标准访问控制列表相比，扩展访问控制列表所检查的数据包元素要丰富很多，它不仅可以检查数据流的源 IP 地址，还可以检查目标 IP 地址、源端口地址、目标端口地址及协议类型。扩展 ACL 通常用于那些需要精确控制的高级的访问。FTP 服务通常使用 TCP 协议的 20 和 21 端口，使用扩展 ACL 可以精确匹配那些访问 FTP 服务的数据包并采取措施。

编号的扩展 ACL 的范围为 100～199 和 2000～2699。
扩展 ACL 通常用在尽量靠近数据流源地址的地方。
定义扩展 ACL 的语法如下。

access-list<num> {deny | permit} icmp {{<sIpAddr><sMask>} | any-source | {host-source <sIpAddr>}} {{<dIpAddr><dMask>} | any-destination | {host-destination <dIpAddr>}} [<icmp-type> [<icmp-code>]] [precedence <prec>] [tos<tos>][time-range<time-range-name>]

任务实现

步骤 1：交换机全部恢复为出厂设置，此处略。

步骤2：配置交换机 A 的 VLAN 信息。

```
DCRS-5650-A#conf
DCRS-5650-A(Config)#vlan 10
DCRS-5650-A(Config-Vlan10)#switchport interface ethernet 0/0/1-8
Set the port Ethernet0/0/1 access vlan 10 successfully
Set the port Ethernet0/0/2 access vlan 10 successfully
Set the port Ethernet0/0/3 access vlan 10 successfully
Set the port Ethernet0/0/4 access vlan 10 successfully
Set the port Ethernet0/0/5 access vlan 10 successfully
Set the port Ethernet0/0/6 access vlan 10 successfully
Set the port Ethernet0/0/7 access vlan 10 successfully
Set the port Ethernet0/0/8 access vlan 10 successfully
DCRS-5650-A(Config-Vlan10)#exit
DCRS-5650-A(Config)#vlan 20
DCRS-5650-A(Config-Vlan20)#switchport interface ethernet 0/0/9-16
Set the port Ethernet0/0/9 access vlan 20 successfully
Set the port Ethernet0/0/10 access vlan 20 successfully
Set the port Ethernet0/0/11 access vlan 20 successfully
Set the port Ethernet0/0/12 access vlan 20 successfully
Set the port Ethernet0/0/13 access vlan 20 successfully
Set the port Ethernet0/0/14 access vlan 20 successfully
Set the port Ethernet0/0/15 access vlan 20 successfully
Set the port Ethernet0/0/16 access vlan 20 successfully
DCRS-5650-A(Config-Vlan20)#exit
DCRS-5650-A(Config)#vlan 100
DCRS-5650-A(Config-Vlan100)#switchport interface ethernet 0/0/24
Set the port Ethernet0/0/24 access vlan 100 successfully
DCRS-5650-A(Config-Vlan100)#exit
DCRS-5650-A(Config)#
```

步骤3：验证交换机 A 的 VLAN 配置。

```
DCRS-5650-A#show vlan
VLAN Name           Type      Media    Ports
-----------------------------------------------------------------
1    default        Static    ENET     Ethernet0/0/17    Ethernet0/0/18
                                       Ethernet0/0/19    Ethernet0/0/20
                                       Ethernet0/0/21    Ethernet0/0/22
                                       Ethernet0/0/23    Ethernet0/0/25
                                       Ethernet0/0/26    Ethernet0/0/27
                                       Ethernet0/0/28
10   VLAN0010       Static    ENET     Ethernet0/0/1     Ethernet0/0/2
                                       Ethernet0/0/3     Ethernet0/0/4
                                       Ethernet0/0/5     Ethernet0/0/6
                                       Ethernet0/0/7     Ethernet0/0/8
20   VLAN0020       Static    ENET     Ethernet0/0/9     Ethernet0/0/10
```

				Ethernet0/0/11	Ethernet0/0/12
				Ethernet0/0/13	Ethernet0/0/14
				Ethernet0/0/15	Ethernet0/0/16
100	VLAN0100	Static	ENET	Ethernet0/0/24	

步骤 4：配置交换机 B 的 VLAN 信息。

```
DCRS-5650-B(Config)#vlan 30
DCRS-5650-B(Config-Vlan30)#switchport interface ethernet 0/0/1-8
Set the port Ethernet0/0/1 access vlan 30 successfully
Set the port Ethernet0/0/2 access vlan 30 successfully
Set the port Ethernet0/0/3 access vlan 30 successfully
Set the port Ethernet0/0/4 access vlan 30 successfully
Set the port Ethernet0/0/5 access vlan 30 successfully
Set the port Ethernet0/0/6 access vlan 30 successfully
Set the port Ethernet0/0/7 access vlan 30 successfully
Set the port Ethernet0/0/8 access vlan 30 successfully
DCRS-5650-B(Config-Vlan30)#exit
DCRS-5650-B(Config)#vlan 40
DCRS-5650-B(Config-Vlan40)#switchport interface ethernet 0/0/9-16
Set the port Ethernet0/0/9 access vlan 40 successfully
Set the port Ethernet0/0/10 access vlan 40 successfully
Set the port Ethernet0/0/11 access vlan 40 successfully
Set the port Ethernet0/0/12 access vlan 40 successfully
Set the port Ethernet0/0/13 access vlan 40 successfully
Set the port Ethernet0/0/14 access vlan 40 successfully
Set the port Ethernet0/0/15 access vlan 40 successfully
Set the port Ethernet0/0/16 access vlan 40 successfully
DCRS-5650-B(Config-Vlan40)#exit
DCRS-5650-B(Config)#vlan 101
DCRS-5650-B(Config-Vlan101)#switchport interface ethernet 0/0/24
Set the port Ethernet0/0/24 access vlan 101 successfully
DCRS-5650-B(Config-Vlan101)#exit
DCRS-5650-B(Config)#
```

步骤 5：验证交换机 B 的 VLAN 配置。

DCRS-5650-B#show vlan

VLAN	Name	Type	Media	Ports	
1	default	Static	ENET	Ethernet0/0/17	Ethernet0/0/18
				Ethernet0/0/19	Ethernet0/0/20
				Ethernet0/0/21	Ethernet0/0/22
				Ethernet0/0/23	Ethernet0/0/25
				Ethernet0/0/26	Ethernet0/0/27
				Ethernet0/0/28	
10	VLAN0010	Static	ENET	Ethernet0/0/1	Ethernet0/0/2
				Ethernet0/0/3	Ethernet0/0/4

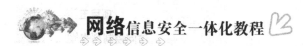

				Ethernet0/0/5	Ethernet0/0/6
				Ethernet0/0/7	Ethernet0/0/8
20	VLAN0020	Static	ENET	Ethernet0/0/9	Ethernet0/0/10
				Ethernet0/0/11	Ethernet0/0/12
				Ethernet0/0/13	Ethernet0/0/14
				Ethernet0/0/15	Ethernet0/0/16
100	VLAN0100	Static	ENET	Ethernet0/0/24	

步骤6：配置交换机 A 的 VLAN 虚拟接口的 IP 地址。

```
DCRS-5650-A(Config)#int vlan 10
DCRS-5650-A(Config-If-Vlan10)#ip address 192.168.10.1 255.255.255.0
DCRS-5650-A(Config-If-Vlan10)#no shut
DCRS-5650-A(Config-If-Vlan10)#exit
DCRS-5650-A(Config)#int vlan 20
DCRS-5650-A(Config-If-Vlan20)#ip address 192.168.20.1 255.255.255.0
DCRS-5650-A(Config-If-Vlan20)#no shut
DCRS-5650-A(Config-If-Vlan20)#exit
DCRS-5650-A(Config)#int vlan 100
DCRS-5650-A(Config-If-Vlan100)#ip address 192.168.100.1 255.255.255.0
DCRS-5650-A(Config-If-Vlan100)#no shut
DCRS-5650-A(Config-If-Vlan100)#exit
DCRS-5650-A(Config)#
```

步骤7：配置交换机 B 的 VLAN 虚拟接口的 IP 地址。

```
DCRS-5650-B(Config)#int vlan 30
DCRS-5650-B(Config-If-Vlan30)#ip address 192.168.30.1 255.255.255.0
DCRS-5650-B(Config-If-Vlan30)#no shut
DCRS-5650-B(Config-If-Vlan30)#exit
DCRS-5650-B(Config)#int vlan 101
DCRS-5650-B(Config-If-Vlan101)#ip address 192.168.100.2 255.255.255.0
DCRS-5650-B(Config-If-Vlan101)#exit
DCRS-5650-B(Config)#
```

步骤8：配置静态路由。

```
DCRS-5650-A(Config)#ip route 0.0.0.0 0.0.0.0 192.168.100.2
DCRS-5650-B(Config)#ip route 0.0.0.0 0.0.0.0 192.168.100.1
```

步骤9：查看交换机 A 的路由表。

```
DCRS-5650-A#show ip route
Codes: K - kernel, C - connected, S - static, R - RIP, B - BGP
       O - OSPF, IA - OSPF inter area
       N1 - OSPF NSSA external type 1, N2 - OSPF NSSA external type 2
       E1 - OSPF external type 1, E2 - OSPF external type 2
       i - IS-IS, L1 - IS-IS level-1, L2 - IS-IS level-2, ia - IS-IS inter area
         * - candidate default
Gateway of last resort is 192.168.100.2 to network 0.0.0.0
S*       0.0.0.0/0 [1/0] via 192.168.100.2, Vlan100
C        127.0.0.0/8 is directly connected, Loopback
```

C	192.168.10.0/24 is directly connected, Vlan10	
C	192.168.20.0/24 is directly connected, Vlan10	
C	192.168.100.0/24 is directly connected, Vlan100	

步骤 10：在 VLAN 30 端口上配置端口的环回测试功能，保证 VLAN 30 可以连通。

```
DCRS-5650-B(Config)# interface ethernet 0/0/1//任意一个 VLAN 30 内的接口均可
DCRS-5650-B(Config-If-Ethernet0/0/1)#loopback
DCRS-5650-B(Config-If-Ethernet0/0/1)#no shut
DCRS-5650-B(Config-If-Ethernet0/0/1)#exit
```

步骤 11：不配置 ACL 验证实验。验证 PC1 和 PC2 是否可以连通 192.168.30.1。

步骤 12：配置扩展 ACL。

```
DCRS-5650-A(Config)#ip access-list extended test2
DCRS-5650-A(Config-Ext-Nacl-test2)#deny icmp 192.168.20.0 0.0.0.255 192.168.30.0 0.0.0.255
//拒绝 192.168.20.0/24 传送数据
DCRS-5650-A(Config-Ext-Nacl-test2)#exit
DCRS-5650-A(Config)#firewall enable                    //启用访问控制列表功能
DCRS-5650-A(Config)#firewall default permit            //默认动作为全部允许通过
DCRS-5650-A(Config)#interface ethernet 0/0/9           //绑定 ACL 到端口上
DCRS-5650-A(Config-Ethernet0/0/9)#ip access-group test2 in
```

步骤 13：验证实验，见表 5-2-4。

表 5-2-4　验证结果

PC	端口	ping	结果	原因
PC1：192.168.10.11/24	0/0/1	192.168.30.1	通	
PC2：192.168.20.11/24	0/0/9	192.168.30.1	不通	

任务三　使用 ACL 过滤特定病毒报文

♂ 需求分析

某公司的网络管理员上班的时候发现公司的网络不正常，经过分析之后发现是公司的计算机中了病毒，使用杀毒软件查杀后发现是"冲击波"和"震荡波"病毒，于是网络管理员想在三层交换机通过配置来隔离这些病毒。

♂ 方案设计

"冲击波"、"震荡波"病毒曾经给网络带来过沉重的打击，到目前为止，Internet 中还有这种病毒及病毒的变种，它们无孔不入，伺机发作。因此，在配置网络设备的时候，采用 ACL 进行过滤，可把这些病毒"拒之门外"，保证网络的稳定运行。

过滤"冲击波"及"冲击波"变种病毒的配置：关闭 TCP 端口 135、139、445 和 593，关闭 UDP 端口 69（TFTP）、135、137 和 138，关闭用于远程命令外壳程序的 TCP 端口 4444。

过滤"震荡波"病毒的配置：关闭 TCP 端口 5554、445、9996。

过滤"SQL 蠕虫"病毒的配置：关闭 TCP 端口 1433，UDP 端口 1434。
所需设备如图 5-3-1 所示。
（1）DCRS-5650 交换机 1 台。
（2）PC 1 台。
（3）Console 线 1 条。
（4）直通网线若干。

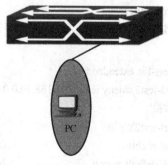

图 5-3-1　使用 ACL 过滤特定病毒报文

任务实现

步骤 1：交换机恢复为出厂设置，此处略。
步骤 2：配置 ACL。

```
Switch(Config)#access-list 110 deny tcp any any d-port 445
Switch(Config)#access-list 110 deny tcp any any d-port 4444
Switch(Config)#access-list 110 deny tcp any any d-port 5554
Switch(Config)#access-list 110 deny tcp any any d-port 9996
Switch(Config)#access-list 110 deny tcp any any d-port 1433
Switch(Config)#access-list 110 deny udp any any d-port 1434
```

步骤 3：启用 ACL。

```
Switch(Config)#firewall enable            //启用访问控制列表功能
Switch(Config)#firewall default permit    //默认动作为全部允许通过
```

步骤 4：应用 ACL。

```
Switch(Config)#interface ethernet0/0/10   //绑定 ACL 到各端口上
Switch(Config-Ethernet0/0/10)#ip access-group 110 in
```

认证考核

选择题

1. 访问控制列表配置中，操作符"gtportnumber"表示控制的是（　　）。
    A．端口号小于此数字的服务
    B．端口号大于此数字的服务
    C．端口号等于此数字的服务

D. 端口号不等于此数字的服务

2．某台路由器上配置了如下访问控制列表：access-list 4 deny 202.38.0.0 0.0.255.255 access-list 4 permit 202.38.160.1 0.0.0.255。这表示（　　）。

A. 只禁止源地址为 202.38.0.0 网段的所有访问
B. 只允许目的地址为 202.38.0.0 网段的所有访问
C. 检查源 IP 地址，禁止 202.38.0.0 大网段的主机访问，但允许其中的 202.38.160.0 小网段上的主机访问
D. 检查目的 IP 地址，禁止 202.38.0.0 大网段的主机访问，但允许其中的 202.38.160.0 小网段的主机访问

3．小于（　　）的端口号已保留并与现有服务一一对应，此数字以上的端口号可自由分配。

A. 100　　　　B. 199　　　　C. 1024　　　　D. 2048

4．下列情况可以使用访问控制列表准确描述的是（　　）。

A. 禁止有 CIH 病毒的文件到主机中
B. 只允许系统管理员访问主机
C. 禁止所有使用 Telnet 的用户访问主机
D. 禁止使用 UNIX 操作系统的用户访问主机

5．如下访问控制列表的含义是（　　）。

access-list 100 deny icmp 10.1.10.10 0.0.255.255 any host-unreachable

A. 规则序列号是 100，禁止到 10.1.10.10 主机的所有主机不可达报文
B. 规则序列号是 100，禁止到 10.1.0.0/16 网段的所有主机不可达报文
C. 规则序列号是 100，禁止从 10.1.0.0/16 网段传送来的所有主机不可达报文
D. 规则序列号是 100，禁止从 10.1.10.10 主机传送来的所有主机不可达报文

6．如下访问控制列表的含义是（　　）。

access-list 102 deny udp 129.9.8.10 0.0.0.255 202.38.160.10 0.0.0.255 gt 128

A. 规则序列号是 102，禁止从 202.38.160.0/24 网段的主机到 129.9.8.0/24 网段的主机使用端口大于 128 的 UDP 进行连接
B. 规则序列号是 102，禁止从 202.38.160.0/24 网段的主机到 129.9.8.0/24 网段的主机使用端口小于 128 的 UDP 进行连接
C. 规则序列号是 102，禁止从 129.9.8.0/24 网段的主机到 202.38.160.0/24 网段的主机使用端口小于 128 的 UDP 进行连接
D. 规则序列号是 102，禁止从 129.9.8.0/24 网段的主机到 202.38.160.0/24 网段的主机使用端口大于 128 的 UDP 进行连接

7．如果在一个接口上使用了 access group 命令，但没有创建相应的 access list，则在此接口上下列描述正确的是（　　）。

A. 发生错误
B. 拒绝所有的数据包 in
C. 拒绝所有的数据包 out
D. 拒绝所有的数据包 in、out
E. 允许所有的数据包 in、out

8. 在访问控制列表中 IP 地址和反掩码为 168.18.64.0 和 0.0.3.255，则表示的 IP 地址范围是（　　）。
   A. 168.18.67.0～168.18.70.255
   B. 168.18.64.0～168.18.67.255
   C. 168.18.63.0～168.18.64.255
   D. 168.18.64.255～168.18.67.255

9. 标准访问控制列表的数字标识范围是（　　）。
   A. 1～50   B. 1～99
   C. 1～100   D. 1～199

10. 标准访问控制列表以（　　）作为判别条件。
    A. 数据包的大小
    B. 数据包的源地址
    C. 数据包的端口号
    D. 数据包的目的地址

11. 访问控制列表 access-list 100 deny ip 10.1.10.10 0.0.255.255 any eq 80 的含义是（　　）。
    A. 规则序列号是 100，禁止到 10.1.10.10 主机的 Telnet 访问
    B. 规则序列号是 100，禁止到 10.1.0.0/16 网段的 WWW 访问
    C. 规则序列号是 100，禁止从 10.1.0.0/16 网段来的 WWW 访问
    D. 规则序列号是 100，禁止从 10.1.10.10 主机来的 rlogin 访问

12. 下列说法正确的是（　　）。
    A. 允许源地址小于 153.19.0.128 的数据包通过
    B. 允许目的地址小于 153.19.0.128 的数据包通过
    C. 允许源地址大于 153.19.0.128、小于 153.19.0.255 的数据包通过
    D. 允许目的地址大于 153.19.0.128 的数据包通过
    E. 配置命令是非法的

13. IP 扩展访问控制列表的数字标识范围是（　　）。
    A. 0～99   B. 1～99   C. 100～199   D. 101～200

14. 使配置的访问控制列表应用到接口上的命令是（　　）。
    A. access-group   B. access-list   C. ip access-list   D. ip access-group

项目六 DHCP 与 VRRP

教学背景

随着企业经营规模的不断扩大，网络应用逐渐增多，原有的网络可能越来越不适应新形式的要求。为此，一些企事业单位对网络的安全性、可管理性和稳定性有了更高的需求。例如，为了保证网络出口的稳定可靠，需要有两条出口线路做冗余备份和负载均衡；为了保证网络安全可靠，要求核心和接入设备支持 VLAN 划分，降低网络内广播数据包的传播，提高带宽资源利用率；网络设备要能够支持灵活多样的管理方式，以减轻管理维护的难度等。

任务一 DHCP 服务器的配置

需求分析

某公司总经理发现自己的计算机发生了"IP 地址冲突"，并且连不上网，于是找来网络管理员解决该问题，网络管理员认为是有员工擅自修改了 IP 地址导致的，通过可以在现有的三层交换机上使用 DHCP 技术来解决该问题。

方案设计

大型网络一般采用 DHCP 作为地址分配的方法，并需要为网络购置多台 DHCP 服务器并放置在网络的不同位置。为减轻网络管理员和用户的配置负担，可以将支持 DHCP 的交换机配置成 DHCP 服务器。

所需设备如图 6-1-1 所示。

（1）DCRS-5650 交换机 1 台。
（2）PC 1~3 台。
（3）Console 线 1 条。
（4）直通网线若干。

为处于不同 VLAN 的 PC 设置 DHCP 服务器，在交换机上划分两个基于端口的 VLAN——VLAN 10 和 VLAN 20，并划分端口和设置 IP 地址，见表 6-1-1。交换机上配置的地址池见表 6-1-2。

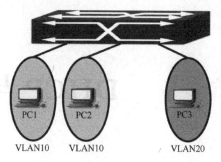

图 6-1-1　DHCP 服务器的配置

表 6-1-1　VLAN 信息表

VLAN	IP 地址	端口成员
10	192.168.10.1/24	1～8
20	192.168.20.1/24	9～16

表 6-1-2　配置地址池

地址池 A　(network 192.168.10.0/24)		地址池 B　(network 192.168.20.0/24)	
设备	IP 地址	设备	IP 地址
默认网关	192.168.10.1	默认网关	192.168.20.1
DNS 服务器	192.168.1.1	DNS 服务器	192.168.1.1
租期	8 小时	租期	1 小时

其中，在 VLAN 10 处，因为工作的需要，特地给一台 MAC 地址为 **00-A0-D1-D1-07-FF** 的机器分配了固定的 IP 地址 192.168.10.88。

知识准备

DHCP 是由服务器控制的一段 IP 地址，客户机登录服务器时可以自动获得服务器分配的 IP 地址、子网掩码、网关和 DNS 地址。首先，网络中必须存在一台 DHCP 服务器，这台服务器是采用 Windows Server 2003 操作系统的计算机，也可以是交换机或路由器设备；其次，客户机计算机要设置为自动获取 IP 地址才能正常获取到 DHCP 服务器提供的 IP 地址。其工作过程如下。

（1）启动 DHCP 服务器。

（2）DHCP 客户机连接在交换机上，但无法获得 IP 地址。遇到这种情况时可检测 DHCP Server 内是否有与交换机 VLAN 接口在同一个网段的地址池，如没有，则请添加该网段的地址池。

（3）在 DHCP 服务中，动态分配 IP 地址与手工分配 IP 地址的地址池是互斥的，即在一个地址池中执行命令 network 和 host 时，只能有一个生效；在手工地址池中，一个地址池内只能配置一对 IP-MAC 的绑定，如果需要建立多对绑定，则可以建立多个手工地址池，在每个地址池中分别配置 IP-MAC 的绑定，否则在同一地址池内新的配置会覆盖旧的配置。

任务实现

步骤 1：交换机全部恢复为出厂设置，此处略。

项目六 DHCP 与 VRRP

步骤 2：创建 VLAN 10 和 VLAN 20，并将相应端口加入 VLAN。

switch(Config)#
switch(Config)#vlan 10
switch(Config-Vlan10)#switchport interface ethernet 0/0/1-8
//给 VLAN 10 加入端口 1~8
Set the port Ethernet0/0/1 access vlan 100 successfully
Set the port Ethernet0/0/2 access vlan 100 successfully
Set the port Ethernet0/0/3 access vlan 100 successfully
Set the port Ethernet0/0/4 access vlan 100 successfully
Set the port Ethernet0/0/5 access vlan 100 successfully
Set the port Ethernet0/0/6 access vlan 100 successfully
Set the port Ethernet0/0/7 access vlan 100 successfully
Set the port Ethernet0/0/8 access vlan 100 successfully
switch(Config-Vlan10)#exit
switch(Config)#vlan 20
switch(Config-Vlan20)#switchport interface ethernet 0/0/9-16
//给 VLAN 20 加入端口 9~16
Set the port Ethernet0/0/9 access vlan 200 successfully
Set the port Ethernet0/0/10 access vlan 200 successfully
Set the port Ethernet0/0/11 access vlan 200 successfully
Set the port Ethernet0/0/12 access vlan 200 successfully
Set the port Ethernet0/0/13 access vlan 200 successfully
Set the port Ethernet0/0/14 access vlan 200 successfully
Set the port Ethernet0/0/15 access vlan 200 successfully
Set the port Ethernet0/0/16 access vlan 200 successfully
switch(Config-Vlan20)#exit
switch(Config)#

步骤 3：为交换机设置 IP 地址。

switch(Config)#interface vlan 1
switch(Config-If-Vlan1)#ip address 192.168.1.1 255.255.255.0
switch(Config-If-Vlan1)#no shutdown
switch(Config)#interface vlan 10
switch(Config-If-Vlan10)#ip address 192.168.10.1 255.255.255.0
switch(Config-If-Vlan100)#no shutdown
switch(Config)#interface vlan20
switch(Config-If-Vlan20)#ip address 192.168.20.1 255.255.255.0
switch(Config-If-Vlan20)#no shutdown

步骤 4：配置 DHCP 服务器。

switch(Config)#service dhcp   //启用 DHCP 服务
switch(Config)#ip dhcp pool testA   //定义地址池
switch(dhcp-testA-config)#network-address 192.168.10.0 24
switch(dhcp-testA-config)#lease 0 8
switch(dhcp-testA-config)#default-router 192.168.10.1
switch(dhcp-testA-config)#dns-server 192.168.1.1

```
switch(dhcp-testA-config)#exit
switch(Config)#ip dhcp pool testB
switch(dhcp-testB-config)#network-address 192.168.20.0 24
switch(dhcp-testB-config)#lease 0 1
switch(dhcp-testB-config)#default-router 192.168.20.1
switch(dhcp-testB-config)#dns-server 192.168.1.1
switch(dhcp-testB-config)#exit
switch(Config)#
```

步骤 5：验证配置。使用"ipconfig/renew"命令在 PC 的命令行窗口中检查是否得到了正确的 IP 地址，见表 6-1-3。

表 6-1-3 验证结果

设备	位置	动作	结果
PC1	1～8 端口	ipconfig/renew	192.168.10.2/24
PC2	1～8 端口	ipconfig/renew	192.168.10.3/24
PC3	9～16 端口	ipconfig/renew	192.168.20.2/24

步骤 6：为特殊的 PC 配置地址池。

```
switch(Config)#ip dhcp excluded-address 192.168.10.77 192.168.10.99
//排除地址池中的不用于动态分配的地址
switch(Config)#ip dhcp pool testC
switch(dhcp-testC-config)#host 192.168.100.88
//手工绑定地址时，分配给指定客户机的用户的 IP 地址
switch(dhcp-testC-config)#hardware-address 00-a0-d1-d1-07-ff
//手工分配地址时，指定用户的硬件地址
switch(dhcp-testC-config)#default-router 192.168.10.1
switch(dhcp-testC-config)#exit
```

步骤 7：验证实验。使用"ipconfig/renew"命令在 PC 的命令行窗口中检查是否得到了正确的 IP 地址，见表 6-1-4。

表 6-1-4 验证结果

设备	位置	动作	结果
PC1	1～8 端口	ipconfig/renew	192.168.100.88/24
PC2	1～8 端口	ipconfig/renew	192.168.100.3/24
PC3	9～16 端口	ipconfig/renew	192.168.200.2/24

任务二 DHCP 中继功能的配置

需求分析

某公司的计算机在同一个网络中，但不在同一个 VLAN 中，现网络中有一台 DHCP 服务

器,负责全网络的动态主机 IP 地址分配,在现有网络不改变的情况下,网络管理员决定利用 DHCP 中继代理来实现。

♂ 方案设计

当 DHCP 客户机和 DHCP 服务器不在同一个网段中时,由 DHCP 中继传递 DHCP 报文。增加 DHCP 中继功能的好处是不必为每个网段都设置 DHCP 服务器,同一个 DHCP 服务器可以为多个子网的客户机提供网络配置参数,这样既节约了成本,又方便了管理。这就是 DHCP 中继的功能。

所需设备如图 6-2-1 所示。
(1) DCRS-5650 交换机 2 台。
(2) PC 3 台。
(3) Console 线 1 条。
(4) 直通网线若干。

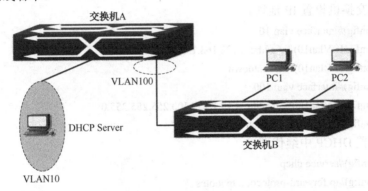

图 6-2-1 DHCP 中继功能的配置

在交换机 A 上划分两个基于端口的 VLAN——VLAN 10 和 VLAN 100,并加入端口和配置 IP 地址,见表 6-2-1。交换机 A 的端口 24 连接一台 DHCP 服务器,服务器的 IP 地址为 10.1.157.1/24。交换机 A 的端口 2 连接交换机 B 的端口 24,交换机 B 不做任何配置,当作集线器来用。DHCP 服务器的 IP 地址池中的地址范围为 192.168.1.10/24~192.168.1.100/24。

表 6-2-1 交换机 A 的 VLAN 信息

VLAN	IP 地址	端口成员
10	192.168.1.1/24	1
100	10.1.157.100/24	24

♂ 知识准备

在大型的网络中,可能存在多个子网。DHCP 客户机通过网络广播消息获得 DHCP 服务器的响应后得到 IP 地址。但广播消息是不能跨越子网的。因此,如果 DHCP 客户机和服务器在不同的子网内,客户机还能不能向服务器申请 IP 地址呢?这就要用到 DHCP 中继代理。DHCP 中继代理实际上是一种软件技术,安装了 DHCP 中继代理的计算机被称为 DHCP 中继代理服务器,它承担不同子网间的 DHCP 客户机和服务器的通信任务。

中继代理是在不同子网上的客户端和服务器之间中转 DHCP/BOOTP 消息的小程序。根据征求意见文档（RFC），DHCP/BOOTP 中继代理是 DHCP 和 BOOTP 标准及功能的一部分。

若中间负责转发 DHCP 报文的交换机、路由器不具备 DHCP 中继功能，则建议替换掉中间的设备或更新版本，使其具备 DHCP 中继功能，才能正确使用这个功能。

任务实现

步骤 1：交换机全部恢复为出厂设置，此处略。

步骤 2：创建 VLAN 10 和 VLAN 100，并将端口加入相应 VLAN。

```
switch(Config)#
switch(Config)#vlan 10
switch(Config-Vlan10)#switchport interface Ethernet0/0/1
switch(Config-Vlan10)#exit
switch(Config)#vlan 100
switch(Config-Vlan100)#switchport interface Ethernet0/0/24
switch(Config-Vlan100)#exit
```

步骤 3：为交换机设置 IP 地址。

```
switch(Config)#interface vlan 10
switch(Config-If-Vlan10)#ip address 192.168.1.1 255.255.255.0
switch(Config-If-Vlan10)#no shutdown
switch(Config)#interface vlan 100
switch(Config-If-Vlan100)#ip address 10.1.157.100 255.255.255.0
switch(Config-If-Vlan100)#no shutdown
```

步骤 4：配置 DHCP 中继代理。

```
switch(Config)#service dhcp
switch(Config)#ip forward-protocol udpbootps
switch(Config)#interface vlan 10
switch(Config-If-Vlan10)#ip helper-address 10.1.157.1
switch(Config-If-Vlan10)#exit
switch(Config)#
```

步骤 5：验证实验。使用"ipconfig/renew"命令在 PC 的命令行窗口中检查是否得到了正确的 IP 地址。

任务三　实现 VRRP 配置

需求分析

某公司企业网络核心层原来采用了一台三层交换机，随着网络应用的日益增多，对网络的可靠性提出了越来越高的要求，公司决定采用默认网关进行冗余备份，以便在其中一台设备出现故障时，备份设备能够及时接管数据转发工作，为用户提供透明的切换，提高网络的稳定性。

方案设计

VRRP（Virtual Router Redundancy Protocol，虚拟路由器冗余协议）是由 IETF 提出的，是

一种标准协议。

VRRP 是一种容错协议，运行于局域网的多台路由器上，它将几台路由器组织成一台"虚拟"路由器，或称为一个备份组。在 VRRP 备份组内，总有一台路由器或以太网交换机是活动路由器，它完成"虚拟"路由器的工作；该备份组中其他的路由器或以太网交换机作为备份路由器（可以不只一台）使用，随时监控活动路由器的活动。当原有的活动路由器出现故障时，各备份路由器将自动选举出一个新的活动路由器来接替其工作，继续为网段内各主机提供路由服务。由于这个选举和接替阶段短暂而平滑，因此，网段内各主机仍然可以正常地使用虚拟路由器，以不间断地与外界保持通信。

所需设备如图 6-3-1 所示。
（1）DCRS-5650 交换机 2 台。
（2）交换机 2 台。
（3）PC 2~4 台。
（4）Console 线 1 或 2 条。
（5）直通网线若干根。

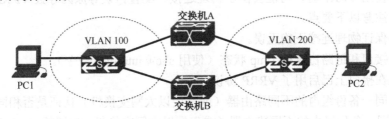

图 6-3-1 实现 VRRP 配置

在交换机 A 和交换机 B 上分别划分基于端口的 VLAN，见表 6-3-1。

表 6-3-1 VLAN 的划分

交换机	VLAN	端口成员	IP 地址
交换机 A	100	1	10.0.0.1/24
	200	24	20.0.0.1/24
交换机 B	100	1	10.0.0.2/24
	200	24	20.0.0.2/24

PC1 和 PC2 的网络设置见表 6-3-2。

表 6-3-2 PC1 和 PC2 的网络设置

设备	IP 地址	网关	子网掩码
PC1	10.0.0.3	10.0.0.10	255.255.255.0
PC2	20.0.0.3	20.0.0.10	255.255.255.0

验证：无论拔掉哪一台三层交换机的线路，PC1 和 PC2 都不需要做网络设置的改变即可与对方进行通信，这证明 VRRP 正常工作。

知识准备

VRRP 是由 IETF 提出的解决局域网中配置静态网关出现单点失效现象的路由协议，1998

年已推出正式的 RFC 2338 协议标准。VRRP 广泛应用在边缘网络中，它的设计目标是支持特定情况下 IP 数据流量失败转移而不会引起混乱，允许主机使用单路由器，以及及时在第一跳路由器使用失败的情形下维护路由器间的连通性。

VRRP 是一种选择协议，它可以把一个虚拟路由器的责任动态地分配到局域网上的 VRRP 路由器中。控制虚拟路由器 IP 地址的 VRRP 路由器称为主路由器，它负责转发数据包到这些虚拟 IP 地址上。一旦主路由器不可用，这种选择过程就提供了动态的故障转移机制，这就允许虚拟路由器的 IP 地址可以作为终端主机的默认第一跳路由器。一个局域网络内的所有主机都设置了默认网关，这样主机发出的目的地址不在本网段的报文将被通过默认网关发往三层交换机，从而实现了主机和外部网络的通信。

VRRP 是一种路由容错协议，也称备份路由协议。一个局域网络内的所有主机都设置了默认路由，当网络内主机发出的目的地址不在本网段时，报文将被通过默认路由发往外部路由器，从而实现了主机与外部网络的通信。当默认路由器端口关闭之后，内部主机将无法与外部通信，如果路由器设置了 VRRP，则虚拟路由将启用备份路由器，从而实现全网通信。

在配置和使用 VRRP 时，可能会由于物理连接、配置错误等原因导致 VRRP 未正常运行。因此，用户应注意以下要点。

（1）应该保证物理连接正确无误。
（2）保证接口和链路协议处于 up 状态（使用 show interface 命令）。
（3）确保在接口上已启用了 VRRP 协议。
（4）检查同一备份组内的不同路由器（或三层以太网交换机）认证是否相同。
（5）检查同一备份组内的不同路由器（或三层以太网交换机）配置的 clock 时间是否相同。
（6）检查虚拟 IP 地址是否和接口真实 IP 地址在同一网段内。

任务实现

步骤 1：交换机全部恢复为出厂设置，此处略。
步骤 2：配置交换机的 VLAN 信息，此处略。
步骤 3：配置交换机各 VLAN 虚拟接口的 IP 地址，此处略。
步骤 4：交换机 A 与交换机 B 互通，此处略。
步骤 5：配置交换机 A 的 VRRP。

```
DCRS-5650-A(config)#router vrrp 1
DCRS-5650-A(config-router)# virtual-ip 10.0.0.10
DCRS-5650-A(config-router)#priority 150
DCRS-5650-A(config-router)# interface vlan 100
DCRS-5650-A(config-router)# enable
DCRS-5650-A(config-router)# exit
DCRS-5650-A(config)#router vrrp2
DCRS-5650-A(config-router)# virtual-ip 20.0.0.10
DCRS-5650-A(config-router)#priority 150
DCRS-5650-A(config-router)# interface vlan 200
DCRS-5650-A(config-router)# enable
DCRS-5650-A(config-router)# exit
```

步骤 6：配置交换机 B 的 VRRP。

```
DCRS-5650-B(config)#router vrrp 1
DCRS-5650-B(config-router)# virtual-ip 10.0.0.10
DCRS-5650-A(config-router)#priority 50
DCRS-5650-A(config-router)#preempt-mode false
/*VRRP 默认为抢占模式,关闭优先级低的交换机 B 的抢占模式以保证高优先级的交换机 A 在故障恢复后,能主动抢占并成为活动路由*/
DCRS-5650-B(config-router)# interface vlan 100
DCRS-5650-B(config-router)# enable
DCRS-5650-B(config-router)# exit
DCRS-5650-B(config)#router vrrp 2
DCRS-5650-B(config-router)# virtual-ip 20.0.0.10
DCRS-5650-A(config-router)#priority 50
DCRS-5650-A(config-router)#preempt-mode false
DCRS-5650-B(config-router)# interface vlan 200
DCRS-5650-B(config-router)# enable
DCRS-5650-B(config-router)# exit
```

步骤 7：在交换机 A 上验证配置。

```
DCRS-5650-A# show vrrp
VrId 1
  State is Master
  Virtual IP is 10.0.0.10 (Not IP owner)
  Interface is Vlan100
  Priority is 150
  Advertisement interval is 1 sec
  Preempt mode is TRUE
VrId2
  State is Master
  Virtual IP is 20.0.0.10 (Not IP owner)
  Interface is Vlan200
  Priority is 150
  Advertisement interval is 1 sec
  Preempt mode is TRUE
```

由此可见：VRRP 已经成功建立，并且活动路由器是路由器 A。

步骤 8：验证实验。在 PC1 上使用"ping 20.0.0.3 -t"命令，并且在过程中拔掉 10.0.0.1 的网线，观察情况。

认证考核

实训题

背景和需求：随着业务的发展，BENET 公司对互联网的访问要求越来越高，因此公司决定采用冗余路由器及两条连接到互联网的链路，以保障到互联网的访问不间断。

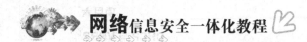

作为网络管理人员，需要设计实施出口路由设备和出口链路的冗余备份。两台三层交换机分别为两个子网提供访问外部网络的出口（两个子网分别为对方的外部网络），并用 VRRP 实现设备级的冗余，以保障网络的连通，如图 6-3-2 所示。

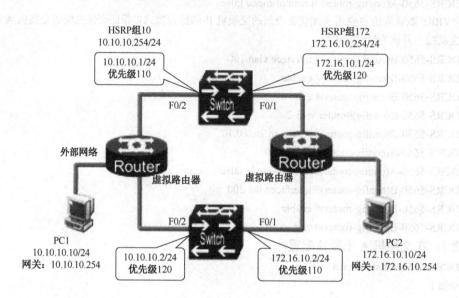

图 6-3-2　网络拓扑

（1）VRRP 的基本配置。

完成标准：使用命令可以看到配置已经生效。

（2）配置 VRRP 的优先级。

完成标准：在两台路由器上使用命令可看到各自的优先级已经配置成功。

（3）配置 VRRP 占先权。

完成标准：在两台路由器上使用命令可看到占先权已配置；LAN 内的 PC 配置 VRRP 虚拟 IP 地址为网关，中断组内任意一台路由器时，与外部通信不中断。

（4）配置 VRRP 的端口跟踪。

完成标准：中断活动路由器的外出接口线路，使用命令可以看到原备份路由器成为活动路由器，两个网段的 PC 之间通信不中断。

项目七 组播

教学背景

组播协议分为主机–路由器之间的组成员关系协议和路由器–路由器之间的组播路由协议。组成员关系协议包括 IGMP（Internet Group Management Protocol，互联网组管理协议）。组播路由协议分为域内组播路由协议及域间组播路由协议。域内组播路由协议包括 PIM-SM、PIM-DM、DVMRP 等，域间组播路由协议包括 MBGP、MSDP 等。

任务一 使用 DVMRP 实现交换机组播的三层对接

需求分析

当信息（包括数据、语音和视频）传送的目的地是网络中的少数用户时，可以采用多种传送方式。例如，可以采用单播（Unicast）的方式，即为每个用户单独建立一条数据传送通路；或者采用广播（Broadcast）的方式，把信息传送给网络中的所有用户，不管它们是否需要，都会接收到广播来的信息。

方案设计

IP 组播技术的出现及时解决了这个问题。组播源仅发送一次信息，组播路由协议为组播数据包建立了树形路由，被传递的信息在尽可能远的分叉路口才开始复制和分发，因此，信息能够被准确高效地传送到每个需要它的用户中。

本任务介绍的是三层环境下的组播对接实验，本任务以目前常用的组播协议 DVMRP 为例进行介绍。DVMRP（Distance Vector Multicast Routing Protocol，距离矢量组播路由协议）是一种密集模式的组播路由协议，采用类似 RIP 方式的路由交换给每个源建立了一个转发广播树，然后通过动态的剪枝/嫁接给每个源建立起一个截断广播树，即到源的最短路径树。通过反向路径检查来决定组播包是否应该被转发到下游。

所需设备如图 7-1-1 所示。

（1）DCRS-5650 交换机 2 台。
（2）PC 2~4 台。
（3）Console 线 1 或 2 条。

（4）直通网线 2~8 条。

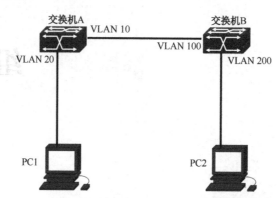

图 7-1-1　实现交换机组播三层对接

任务要求：在交换机上划分基于端口的 VLAN，见表 7-1-1。所有 PC1 是组播服务器，PC2 是客户端，在组播服务器上运行服务器软件 Wsend.exe，在 PC 上运行客户端软件 MCastTest20，查看组播状态。

表 7-1-1　交换机上划分基于端口的 VLAN

交换机	VLAN	端口成员	IP 地址	连接
交换机 A	10	E0/0/11	10.0.0.1/24	交换机 B E0/0/24
交换机 A	20	E0/0/1	20.0.0.1/24	20.0.0.2/24
交换机 B	100	E0/0/11	10.0.0.2/24	交换机 A E0/0/11
交换机 B	200	E0/0/1	30.0.0.1/24	30.0.0.2/24

♂ 知识准备

DVMRP 是一种互联网路由协议，为互联网络的主机组提供了一种面向无连接信息组播的有效机制。DVMRP 是一种"内部网关协议"，适合在自治系统内使用，不适合在不同的自治系统之间使用。当前开发的 DVMRP 不能用于为非组播数据报路由，因此要想使一个路由器既能为多播数据报路由，又能为单播数据报路由，它必须运行两个不同的路由选择进程。DVMRP 数据包封装于 IP 数据报中，使用的 IP 协议号为 2，这与 Internet 组管理协议相同。

DVMRP 的开发基于 RIP。DVMRP 整合了 RIP 中的许多特性和截断方向路径广播算法。另外，为了试验跨越不支持多播的网络的可行性，开发了一种称为"隧道"的机制。DVMRP 和 RIP 的主要不同之处在于：RIP 路由和转发数据报到明确的目的地。DVMRP 的目的是跟踪到组播数据报出发地的返回路径。

DVMRP 具有如下重要特性。

（1）用于决定反向路径检查信息的路由交换，以距离矢量为基础（方式与 RIP 相似）。

（2）路由交换更新周期性地发生（默认为 60s）。

（3）TTL 上限=32 跳（而 RIP 是 16 跳）。

（4）路由更新包括掩码，支持 CIDR。

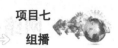

任务实现

步骤1：交换机全部恢复为出厂设置，此处略。

步骤2：配置交换机 A 的 VLAN 信息。

```
DCRS-5650-A(Config)#vlan 10
DCRS-5650-A(Config-Vlan10)#switchport interface ethernet 0/0/11
DCRS-5650-A(Config-Vlan10)#exit
DCRS-5650-A(Config)#int vlan 10
DCRS-5650-A(Config-If-Vlan10)#ip add 10.0.0.1 255.255.255.0
DCRS-5650-A(Config-If-Vlan10)#exit
DCRS-5650-A(Config)#vlan 20
DCRS-5650-A(Config-Vlan20)#switchport interface ethernet 0/0/1
DCRS-5650-A(Config-Vlan10)#exit
DCRS-5650-A(Config)#int vlan 20
DCRS-5650-A(Config-If-Vlan20)#ip add 20.0.0.1 255.255.255.0
DCRS-5650-A(Config-If-Vlan20)#exit
DCRS-5650-A(Config)#
```

步骤3：配置交换机 B 的 VLAN 信息。

```
DCRS-5650-B(Config)#vlan 100
DCRS-5650-B(Config-Vlan100)#switchport interface ethernet 0/0/11
DCRS-5650-B(Config-Vlan100)#exit
DCRS-5650-B(Config)#int vlan 100
DCRS-5650-B(Config-If-Vlan100)#ip add 10.0.0.2 255.255.255.0
DCRS-5650-B(Config-If-Vlan100)#exit
DCRS-5650-B(Config)# vlan 200
DCRS-5650-B(Config-Vlan200)#switchport interface ethernet 0/0/1
DCRS-5650-B(Config-Vlan200)#exit
DCRS-5650-B(Config)#int vlan 200
DCRS-5650-B(Config-If-Vlan200)#ip add 30.0.0.1 255.255.255.0
DCRS-5650-B(Config-If-Vlan200)#exit
DCRS-5650-B(Config)#
```

步骤4：验证配置，在 PC1 上测试其能否连通 PC2，发现不能通信。

步骤5：在交换机 A 上启动 DVMRP 协议。

```
DCRS-5650-A(Config)ip dvmrp multicast-routing    //开启组播协议
DCRS-5650-A(Config)#intvlan 10
DCRS-5650-A(Config-If-Vlan1)#ip dvmrp enable     //在 VLAN 接口上开启 DVMRP 协议
DCRS-5650-A(Config-If-Vlan1)#exit
DCRS-5650-A(Config)#int vlan 20
DCRS-5650-A(Config-If-Vlan20)#ip dvmrp enable
DCRS-5650-A(Config-If-Vlan20)#
DCRS-5650-A(Config-If-Vlan20)#exit
DCRS-5650-A(Config)#
```

步骤6：在交换机 B 上启动 DVMRP 协议。

```
DCRS-5650-B(Config)ip dvmrp multicast-routing
DCRS-5650-B(Config)#int vlan 100
```

```
DCRS-5650-B(Config-If-Vlan100)#ip dvmrp enable
DCRS-5650-B(Config-If-Vlan100)#exit
DCRS-5650-B(Config)#int vlan 200
DCRS-5650-B(Config-If-Vlan200)#ip dvmrp enable
DCRS-5650-B(Config-If-Vlan200)#exit
DCRS-5650-B(Config)#
```

步骤7：验证配置。

```
DCRS-5650-B#show ip dvmrp route
Flags: N = New, D = DirectlyConnected, H = Holddown
Network        Flags NexthopNexthop          Metric Uptime    Exptime
Xface          Neighbor
10.0.0.0/8     .D.   Vlan100   Directly Connected   1    00:00:53 00:00:00
30.0.0.0/8     .D.   Vlan200   Directly Connected   1    00:00:32 00:00:00
DCRS-5650-A#show ip dvmrp neighbor
Neighbor             Interface    Uptime/Expires      Maj  Min  Cap
Address                                               VerVerFlg
10.0.0.2             Vlan10       00:02:00/00:00:25    3   255  2e
RSA#
```

步骤8：组播服务器发送广播包，如图 7-1-2 所示。

步骤9：组播客户端成功接收，如图 7-1-3 所示。

图 7-1-2 服务器发送广播包

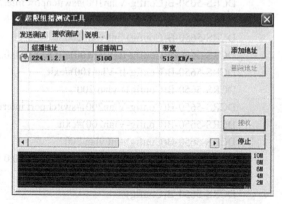

图 7-1-3 客户端接收广播包

PC1 作为组播服务器向 PC2 发送组播报。PC2 作为组播客户端，若可以接收由 PC1 发出的组播报，则证明成功。

由此证明，即使在单播不能够进行通信的情况下，DVMRP 组播协议也能通信，DVMRP 组播协议带有最优路由选择功能。

任务二 使用 PIM 实现交换机组播三层对接

♂ 需求分析

在一个网络上有 200 个用户需要接收相同的信息时，传统的解决方案是用单播方式把这一信息分别发送 200 次，以便确保需要数据的用户能够得到所需的数据；或者采用广播的方式，在整个网络范围内传送数据，需要这些数据的用户可直接在网络中获取。这两种方式都浪费了大量宝贵的带宽资源，而且广播方式也不利于信息的安全和保密。

方案设计

IP 组播技术的出现及时解决了这个问题。组播源仅发送一次信息，组播路由协议为组播数据包建立树形路由，被传递的信息在尽可能远的分叉路口才开始复制和分发，因此，信息能够被准确高效地传送到每个需要它的用户中。

所需设备如图 7-2-1 所示。
（1）DCRS-5650 交换机 2 台。
（2）PC 2～4 台。
（3）Console 线 1 或 2 条。
（4）直通网线 2～8 条。

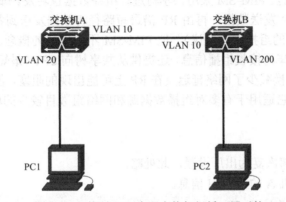

图 7-2-1 使用 PIM 实现交换机组播三层对接

任务要求：在交换机上划分基于端口的 VLAN，见表 7-2-1。PC1 是组播服务器，PC2 是客户端，在组播服务器上运行服务器软件 Wsend.exe，在 PC 上运行客户端软件 MCastTest20，查看组播状态。

表 7-2-1 交换机上划分 VLAN 信息

交换机	VLAN	端口成员	IP 地址	连接
交换机 A	10	E0/0/11	10.0.0.1/24	交换机 B E0/0/24
交换机 A	20	E0/0/1	20.0.0.1/24	20.0.0.2/24
交换机 B	10	E0/0/11	10.0.0.2/24	交换机 A E0/0/11
交换机 B	200	E0/0/1	30.0.0.1/24	30.0.0.2/24

知识准备

PIM（Protocol Independent Multicast，协议无关组播）在 Internet 范围内同时支持 SPT 和共享树，并使两者之间可以灵活转换，因而集中了它们的优点，提高了组播效率。PIM 定义了两种模式：密集模式和稀疏模式。

密集模式通常用于组成员比较密集的网络中。在密集模式中，当有组播源出现时，路由器假设所有的网络都有组成员，只要构建一棵从源开始的转发树，全部网络就都有了组播流量。

稀疏模式通常用于组成员比较稀疏的网络中。在稀疏模式中，路由器假设所有的网络都没有组成员，除非有主机明确表示加入该组。转发树的建立从终端的叶结点组成员开始，扩展到中心的根结点上。

1. PIM-DM

PIM-DM 与 DVMRP 相似，都属于密集模式协议，都采用了"扩散/剪枝"机制。同时，假定带宽不受限制，每个路由器都想接收组播数据包。其不同之处在于 DVMRP 使用内建的组播路由协议，而 PIM-DM 采用 RPF 动态建立 SPT。该模式适用于下述几种情况：高速网络；组播源和接收者比较靠近，发送者少，接收者多；组播数据流比较大且比较稳定。

2. PIM-SM

PIM-SM 与基于"扩散/剪枝"模型的根本差别在于 PIM-SM 基于显式加入模型，即接收者向 RP（Rendezvous Point，汇合点）发送加入消息，而路由器只在已加入某个组播组输出接口上转发该组播组的数据包。PIM-SM 采用共享树进行组播数据包转发。每一个组有一个汇合点，组播源沿最短路径向 RP 发送数据，再由 RP 沿最短路径将数据发送到各个接收端。这一点类似于 CBT（基于核心树的组播路由协议），但 PIM-SM 不使用核的概念。PIM-SM 主要优势之一是它不仅可以通过共享树接收组播信息，还提供从共享树向 SPT（最短路径树）转换的机制。尽管从共享树向 SPT 转换减少了网络延迟及在 RP 上可能出现的阻塞，但这种转换耗费了相当多的路由器资源，所以它适用于有多对组播数据源和网络组数目较少的环境。

任务实现

步骤 1：交换机全部恢复为出厂设置，此处略。

步骤 2：配置交换机 A 的 VLAN 信息。

```
DCRS-5650-A(Config)#vlan 10
DCRS-5650-A(Config-Vlan10)#switchport interface ethernet 0/0/11
DCRS-5650-A(Config-Vlan10)#exit
DCRS-5650-A(Config)#int vlan 10
DCRS-5650-A(Config-If-Vlan10)#ip add 10.0.0.1 255.255.255.0
DCRS-5650-A(Config-If-Vlan10)#exit
DCRS-5650-A(Config)#vlan 20
DCRS-5650-A(Config-Vlan20)#switchport interface ethernet 0/0/1
DCRS-5650-A(Config-Vlan10)#exit
DCRS-5650-A(Config)#int vlan 20
DCRS-5650-A(Config-If-Vlan20)#ip add 20.0.0.1 255.255.255.0
DCRS-5650-A(Config-If-Vlan20)#exit
DCRS-5650-A(Config)#
```

步骤 3：配置交换机 B 的 VLAN 信息。

```
DCRS-5650-B(Config)#vlan 10
DCRS-5650-B(Config-Vlan10)#switchport interface ethernet 0/0/11
DCRS-5650-B(Config-Vlan10)#exit
DCRS-5650-B(Config)#int vlan 10
DCRS-5650-B(Config-If-Vlan10)#ip add 10.0.0.2 255.255.255.0
```

```
DCRS-5650-B(Config-If-Vlan10)#exit
DCRS-5650-B(Config)#vlan 200
DCRS-5650-B(Config-Vlan200)#switchport interface ethernet 0/0/1
DCRS-5650-B(Config-Vlan200)#exit
DCRS-5650-B(Config)#int vlan 200
DCRS-5650-B(Config-If-Vlan200)#ip add 30.0.0.1 255.255.255.0
DCRS-5650-B(Config-If-Vlan200)#exit
DCRS-5650-B(Config)#
```

步骤4：验证配置，PC1 ping PC2，发现不能够通信。

步骤5：在交换机 A 上启动 PIM 协议。

```
DCRS-5650-A(Config)ip pim multicast-routing        //开启组播协议
DCRS-5650-A(Config)#int vlan 10
DCRS-5650-A(Config-If-Vlan1)#ip pim dense-mode     //在 VLAN 接口上开启 PIM 协议
DCRS-5650-A(Config-If-Vlan1)#exit
DCRS-5650-A(Config)#int vlan 20
DCRS-5650-A(Config-If-Vlan20)#ip pim dense-mode
DCRS-5650-A(Config-If-Vlan20)#exit
DCRS-5650-A(Config)#
```

步骤6：在交换机 B 上启动 PIM 协议。

```
DCRS-5650-B(Config)ip pim multicast-routing
DCRS-5650-B(Config)#int vlan 10
DCRS-5650-B(Config-If-Vlan10)#ip pim dense-mode
DCRS-5650-B(Config-If-Vlan10)#exit
DCRS-5650-B(Config)#int vlan 200
DCRS-5650-B(Config-If-Vlan200)#ip pim dense-mode
DCRS-5650-B(Config-If-Vlan200)#exit
DCRS-5650-B(Config)#
```

步骤7：验证配置。

```
DCR-5650-B#show ip pim nei
Neighbor          Interface         Uptime/Expires        Ver      DR
Address                                                            Priority/Mode
10.0.0.1          Vlan10            00:00:19/00:01:43 v2           1 /
DCR-5650-B#show ip pim interface detail
Vlan10 (vif 0):
    Address 10.0.0.2, DR 10.0.0.2
    Hello period 30 seconds, Next Hello in 0 seconds
    Triggered Hello period 5 seconds
    Neighbors:
    10.0.0.1
Vlan200 (vif 1):
    Address 30.0.0.1, DR 30.0.0.1
    Hello period 30 seconds, Next Hello in 12 seconds
    Triggered Hello period 5 seconds
    Neighbors:
```

由 PC1 作为组播服务器，向 PC2 发送组播报，PC2 作为组播客户端，不可以接收由 PC1

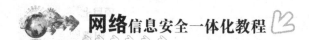

发出的组播报。

步骤 8：由此证明在单播不能够进行通信的情况下，PIM 协议也不能通信，PIM 协议未带有路由选择功能。

步骤 9：配置路由协议。

```
DCRS-5650-A(Config)#ip route 0.0.0.0 0.0.0.0 10.0.0.2
DCRS-5650-A(Config)#ip route 0.0.0.0 0.0.0.0 10.0.0.1
```

步骤 10：验证配置。

由 PC1 ping PC2，发现可以通信。由 PC1 作为组播服务器，向 PC2 发送组播报，PC2 作为组播客户端，可以接收由 PC1 发出的组播报。由此证明 PIM 协议本身不带有最优路径的算法，要依靠单播协议或静态路由完成最优路径的选择，但是并不介意使用的是什么协议。

任务三　交换机组播二层对接

♂ 需求分析

某公司的员工比较多，大概有 200 位用户，经常需要接收相同的信息。网络管理员发现采用单播方式太浪费时间，而采用广播方式又会浪费大量宝贵的带宽资源，也不利于信息的安全和保密，所以网络管理员想采用组播的方式来实现此功能。

♂ 方案设计

通过任务组播三层对接的实验，已经了解了组播的原理和基本配置方法，在本任务中，以 PIM-DM 为例，完成现实生活中使用最为广泛的二层对接环境的模拟。二层对接的 DVMRP、PIM-SM 实验，因教学要求和篇幅所限，考虑到配置的关联性，希望有能力的读者参照前面的任务，在了解协议特点的基础上自行完成。

所需设备如图 7-3-1 所示。

（1）DCRS-5650 交换机 1 台。
（2）DCS-3926S 交换机 1 台。
（3）PC 2~4 台。
（4）Console 线 1 或 2 条。
（5）直通网线 2~8 条。
（6）组播测试软件：发送端、接收端。

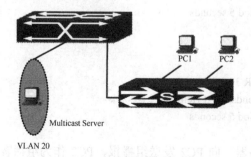

图 7-3-1　交换机组播二层对接拓扑图

任务要求：在交换机上划分基于端口的 VLAN，见表 7-3-1，所有 PC 都是组播客户端，在组播服务器上运行服务器软件 Wsend.exe，在 PC 上运行客户端软件 Wsend.exe，查看组播状态。

表 7-3-1 交换机划分 VLAN 信息

VLAN	端口成员	IP 地址	连接
1	E0/0/24	192.168.10.0/0/24	交换机 E0/0/24
20	E0/0/9	192.168.20.0/0/24	组播服务器

♂ 知识准备

PIM-DM 属于密集模式的组播路由协议，适用于小型网络。在这种网络环境下，组播组的成员相对比较密集。

♂ 任务实现

步骤 1：在三层交换机上恢复出厂设置，此处略。

步骤 2：配置交换机的 VLAN 信息。

```
DCRS-5650(Config)#vlan 20
DCRS-5650(Config-Vlan20)#switchport interface ethernet 0/0/9
Set the port Ethernet0/0/9 access vlan20 successfully
DCRS-5650(Config-Vlan20)#exit
DCRS-5650(Config)#
DCRS-5650(Config)#interface vlan 1
DCRS-5650(Config-If-Vlan1)#ip address 192.168.10.1 255.255.255.0
DCRS-5650(Config-If-Vlan1)#exit
DCRS-5650(Config)#interface vlan 20
DCRS-5650(Config-If-Vlan20)#ip address 192.168.20.1 255.255.255.0
DCRS-5650(Config-If-Vlan20)#exit
```

步骤 3：三层交换机启用 PIM-DM 协议。

```
DCRS-5650 (Config)#ippim multicast-routing        //使能组播协议
DCRS-5650(Config)#int vlan 1
DCRS-5650(Config-If-Vlan1)#ip pim dense-mode      //启用本接口的 PIM-DM 协议
DCRS-5650(Config-If-Vlan1)#exit
DCRS-5650(Config)#int vlan 20
DCRS-5650(Config-If-Vlan20)#ip pim dense-mode
DCRS-5650(Config-If-Vlan20)#exit
DCRS-5650(Config)#
```

步骤 4：验证配置。在组播服务器发送组播数据包，组播客户端可以接收，同时用另外一台计算机开启 Sniffer 进行抓包，可以抓到网络中的组播数据，如图 7-3-2 所示。

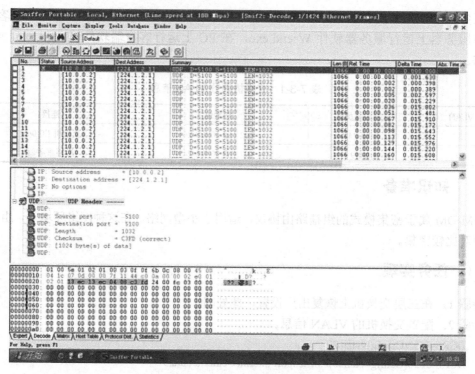

图 7-3-2 抓取到的数据包

步骤 5：二层交换机启用 IGMP 侦听。

在二层交换机上启用指定 VLAN 的 IGMP Snooping 功能，二层交换机默认接收组播数据，可以不做配置，这样组播数据会按照广播的形式传送。

下面以 VLAN 1 为例进行介绍，其他二层功能可参考前面的任务。先配置 VLAN 信息、Trunk 端口等，与三层交换机连通，再进行如下配置。

```
switch(Config)#ip igmp snooping                    //启用 IGMP Snooping 功能
switch(Config)#ip igmp snooping vlan1              /*指定 VLAN 的 IGMP Snooping */ switch(Config)#ip
switch(Config)#ip igmp snooping vlan 1 mroute interface ethernet0/0/2
switch(config)#
```

步骤 6：验证配置。

```
switch#show ip igmp snooping
igmp snooping status                 : Enabled
IGMP information for VLAN 20:
igmp snooping vlan status            :Disabled
igmp snooping vlan query             :Disabled
igmp snooping vlanmrouter port :
--------------------------------
IGMP information for VLAN 1:
igmp snooping vlan status            :Enabled
igmp snooping vlan query             :Disabled
igmp snooping vlanmrouter port :
        Ethernet0/0/2;state :UP
switch#show mac
```

```
Read mac address table....
Vlan Mac Address                  Type              Creator      Ports
--------------------------------------------------------------------------------
1     00-03-0f-0f-6b-0c           DYNAMIC Hardware Ethernet0/0/2
1     00-03-0f-0f-6b-0d           DYNAMIC Hardware Ethernet0/0/2
1     00-26-9e-52-4a-05           DYNAMIC Hardware Ethernet0/0/2
10    00-03-0f-00-a9-59           STATIC  System   CPU
10    00-03-0f-0f-6b-0c           DYNAMIC Hardware Ethernet0/0/2
10    00-0b-cd-4a-97-2e           DYNAMIC Hardware Ethernet0/0/18
10    00-22-64-c0-80-94           DYNAMIC Hardware Ethernet0/0/16
switch#
switch#show mac multicast
Vlan Mac Address                  Type     Creator      Ports
--------------------------------------------------------------------------------
1     01-00-5e-00-00-fc           MULTI    IGMP         Ethernet0/0/2
switch#show mac multicast
Vlan Mac Address                  Type     Creator      Ports
--------------------------------------------------------------------------------
1     01-00-5e-00-00-fc           MULTI    IGMP         Ethernet0/0/2
1     01-00-5e-01-02-01           MULTI    IGMP         Ethernet0/0/2
switch#
```

在二层交换上启用侦听后，在非组播客户端上无法捕获组播客户端上的组播数据，说明组播已经不在以广播的形式发送数据包了。

认证考核

实训题

1. 在三层交换机 SWITCH-X-A 上使用 PIM-SM 方式启用组播功能，使 computer 和 finance 部门之间可以传送组播包，pim hello 报文的发送时间间隔为 60s，在路由器 ROUTER-X-A 上启用组播快速转发功能，组播流的最大输入流量和输出流量都为 10kb/s。

拓扑结构同本项目任务一。

2. 在三层交换机 S3-X-1 上使用 PIM-DM 方式启用组播功能，使 VLAN 7 和 VLAN 8 之间可以传送组播包，在三层交换机 S3-X-2 端口 0/0/21 上配置广播风暴抑制，端口允许通过广播包数为 3000 个/秒。

拓扑结构同本项目任务一。

项目八 广域网技术

♂ 教学背景

WAN（Wide Area Network，广域网）也称远程网，是一种运行地域超过局域网的数据通信网络，通常跨接很大的物理范围，所覆盖的范围从几十千米到几千千米，它能连接多个城市或国家，或者横跨几个大洲并提供远距离通信，形成国际性的远程网络。广域网和局域网的主要区别之一是需要向外部的广域网服务提供商申请并订购广域网电信网络服务。一般使用电信运营商提供的数据链路在广域网范围内访问网络。

任务一 路由器串口 PPP PAP 认证

♂ 需求分析

某公司为了满足不断增长的业务需求，申请了专线接入。公司的路由器与 ISP 进行链路协商时，需要验证身份。配置路由器以保证链路的建立，并考虑其安全性。

♂ 方案设计

WAN 专线链路建立时要进行安全验证，以保证链路的安全性。链路协商时，PAP（Password Authentication Protocol，密码验证协议）在设备之间传输用户名、密码，以实现用户身份的验证确认。公司计划在路由器上配置 PPP PAP 认证，以实现链路的安全连接。

所需设备如图 8-1-1 所示。

（1）DCR-2626 路由器 2 台。
（2）CR-V35MT 1 条。
（3）CR-V35FC 1 条。

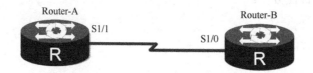

图 8-1-1 路由器串口 PPP PAP 认证

项目八 广域网技术

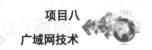

路由器配置信息见表 8-1-1。

表 8-1-1 路由器配置信息

Router-A					Router-B				
接口		IP 地址	账号	密码	接口		IP 地址	账号	密码
S1/1 DCE		192.168.1.1	RouterA	digitalchinaA	S1/0 DTE		192.168.1.2	RouterB	digitalchinaB

知识准备

PPP 支持两种认证方式——PAP 和 CHAP。PAP 是指验证双方通过两次握手完成验证过程，它是一种用于对试图登录到点对点协议服务器上的用户进行身份验证的方法。由被验证方主动发出验证请求，发送的验证包含用户名和密码，由验证方验证后做出回复——通过验证或验证失败。在验证过程中用户名和密码以明文的方式在链路上传输。

PAP 是一种简单的明文验证方式。NAS（Network Access Server，网络接入服务器）要求用户提供用户名和口令，PAP 以明文方式返回用户信息。显然，这种验证方式的安全性较差，第三方可以很容易地获取被传送的用户名和口令，并利用这些信息与 NAS 建立连接获取 NAS 提供的所有资源。所以，一旦用户密码被第三方窃取，PAP 就无法提供避免受到第三方攻击的保障措施了。

任务实现

步骤 1：Router-A 的配置。

Router>enable	//进入特权模式
Router #config	//进入全局配置模式
Router_config#hostname Router-A	//修改机器名
Router-A_config#aaa authentication ppp test local	
Router-A_config#username RouterB password digitalchinaB	//设置账号和密码
Router-A_config#interface s1/0	//进入接口模式
Router-A_config_s1/0#ip address 192.168.1.1 255.255.255.0	//配置 IP 地址
Router-A_config_s1/0#encapsulation PPP	//封装 PPP 协议
Router-A_config_s1/0#ppp authentication pap test	//设置验证方式
Router-A_config_s1/0#ppp pap sent-username RouterA password digitalchinaA	
	//设置发送给对方验证的账号和密码
Router-A_config_s1/0#physical-layer speed 64000	//配置 DCE 时钟频率
Router-A_config_s1/0#no shutdown	
Router-A_config_s1/0#^Z	

步骤 2：查看 Router-A 的配置。

Router-A#show interface s1/1	//查看接口状态
Serial1/0 is up, line protocol is down	//对端没有配置，所以协议处于 DOWN 状态
Mode=Sync DCE Speed=64000	//查看 DCE
DTR=UP,DSR=UP,RTS=UP,CTS=DOWN,DCD=UP	
Interface address is 192.168.1.1/24	//查看 IP 地址
MTU 1500 bytes, BW 64 kbit, DLY 2000 usec	
Encapsulation prototol PPP, link check interval is 10 sec	//查看封装协议

```
        Octets    Received0, Octets Sent 0
        Frames Received 0, Frames Sent 0, Link-check Frames Received0
        Link-check Frames Sent 89,    LoopBack times 0
        Frames Discarded 0, Unknown Protocols Frames Received 0, Sent failuile 0
            Link-check Timeout 0, Queue Error 0, Link Error 0,
            60 second input rate 0 bits/sec, 0 packets/sec!
            60 second output rate 0 bits/sec, 0 packets/sec!
            0 packets input, 0 bytes, 8 unused_rx, 0 no buffer
            0 input errors, 0 CRC, 0 frame, 0 overrun, 0 ignored, 0 abort
            8 packets output, 192 bytes, 0 unused_tx, 0 underruns
error:
            0 clock, 0 grace
        PowerQUICC SCC specific errors:
            0 recvallocbmblk fail      0 recv no buffer
        0 transmitter queue full        0 transmitter hwqueue_full
```

步骤 3：Router-B 的配置。

```
        Router>enable                                          //进入特权模式
        Router #config                                         //进入全局配置模式
        Router_config#hostname Router-B                        //修改机器名
        Router-B_config#aaa authentication ppp test local
        Router-B_config#usernameRouterA password digitalchinaA //设置账号和密码
        Router-B_config#interface s1/0                         //进入接口模式
        Router-B_config_s1/0#ip address 192.168.1.2 255.255.255.0  //配置 IP 地址
        Router-B_config_s1/0#encapsulation PPP                 //封装 PPP 协议
        Router-B_config_s1/0#ppp authentication pap   test     //设置验证方式
        Router-B_config_s1/0#ppp pap sent-username RouterBpassworddigitalchinaB
                                                               //设置发送给对方验证的账号和密码
        Router-B_config_s1/0#shutdown
        Router-B_config_s1/0#no shutdown
        Router-B_config_s1/0#^Z                                //按 Ctrl+Z 组合键进入特权模式
```

步骤 4：查看 Router-B 的配置。

```
        Router-A#show interface s1/0                      //查看接口状态
        Serial1/0 is up, line protocol is up              //接口和协议都处于 up 状态
          Mode=Sync DTE                                   //查看 DTE
            DTR=UP,DSR=UP,RTS=UP,CTS=DOWN,DCD=UP
            Interface address is 192.168.1.2/24           //查看 IP 地址
            MTU 1500 bytes, BW 64 kbit, DLY 2000 usec
            Encapsulation protorol PPP, link check interval is 10 sec   //查看封装协议
        Octets    Received0, Octets Sent 0
        Frames Received 0, Frames Sent 0, Link-check Frames Received0
        Link-check Frames Sent 89,    LoopBack times 0
        Frames Discarded 0, Unknown Protocols Frames Received 0, Sent failuile 0
            Link-check Timeout 0, Queue Error 0, Link Error 0,
            60 second input rate 0 bits/sec, 0 packets/sec!
```

```
            60 second output rate 0 bits/sec, 0 packets/sec!
                0 packets input, 0 bytes, 8 unused_rx, 0 no buffer
                0 input errors, 0 CRC, 0 frame, 0 overrun, 0 ignored, 0 abort
                8 packets output, 192 bytes, 0 unused_tx, 0 underruns
        error:
                0 clock, 0 grace
        PowerQUICC SCC specific errors:
                0 recvallocbmblk fail       0 recv no buffer
                0 transmitter queue full    0 transmitter hwqueue_full
```

步骤 5：测试连通性。

```
Router-A#ping 192.168.1.2
PING 192.168.1.2 (192.168.1.2): 56 data bytes
!!!!!
--- 192.168.1.2 ping statistics ---
5 packets transmitted, 5 packets received, 0% packet loss
round-trip min/avg/max = 20/22/30 ms
```

♂ 小贴士

账号和密码一定要交叉对应，发送的账号和密码要与对方账号数据库中的账号和密码对应；不要忘记配置 DCE 的时钟频率。

任务二　路由器串口 PPP CHAP 认证

♂ 需求分析

考虑到增强 WAN 链路的安全功能，公司的网络管理员计划在路由器上配置 PPP 的 CHAP 认证，以满足日益增长的安全需求。

♂ 方案设计

CHAP（Challenge Hand Authentication Protocol，挑战握手验证协议）使用三次握手机制来启动一条链路和周期性地验证远程结点。与 PAP 相比，CHAP 认证更具有安全性。CHAP 只在网络上传送用户名而不传送口令，因此安全性更高。

所需设备如图 8-1-1 所示。

（1）DCR-2626 路由器 2 台。
（2）CR-V35MT 1 条。
（3）CR-V35FC 1 条。

路由器配置信息如表 8-2-1 所示。

表 8-2-1　路由器配置信息

Router-A				Router-B			
接口	IP 地址	账号	密码	接口	IP 地址	账号	密码
S1/1 DCE	192.168.1.1	RouterA	digitalchinaA	S1/0 DTE	192.168.1.2	RouterB	digitalchinaB

♂ 知识准备

CHAP 是一种加密的验证方式，能够避免建立连接时传送用户的真实密码。NAS 向远程用户发送一个挑战口令，其中包括会话 ID 和一个任意生成的挑战字串（Arbitrary Challenge String）。远程客户必须使用 MD5 单向哈希算法（One-way Hashing Algorithm）返回用户名和加密的挑战口令、会话 ID 及用户口令，其中，用户名以非哈希方式发送。

CHAP 对 PAP 进行了改进，不再直接通过链路发送明文口令，而是使用挑战口令以哈希算法对口令进行加密。因为服务器端存有客户的明文口令，所以服务器可以重复客户端进行的操作，并将结果与用户返回的口令进行对照。CHAP 为每一次验证任意生成一个挑战字符串来防止受到再现攻击（Replay Attack）。在整个连接过程中，CHAP 将不定时地向客户端重复发送挑战口令，从而避免第 3 方冒充远程客户（Remote Client Impersonation）进行攻击。

♂ 任务实现

步骤 1：Router-A 的配置。

```
Router>enable                                              //进入特权模式
Router #config                                             //进入全局配置模式
Router _config#hostname Router-A                           //修改机器名
Router-A_config# aaa authentication ppp test local
//定义一个名为 test，使用本地数据进行验证的 aaa 验证方法
Router-A_config#usernameRouterB password digitalchina      //设置账号和密码
Router-A_config#interface s1/0                             //进入接口模式
Router-A_config_s1/0#ip address 192.168.1.1 255.255.255.0  //配置 IP 地址
Router-A_config_s1/0#encapsulation PPP                     //封装 PPP 协议
Router-A_config_s1/0#ppp authentication chap   test        //设置验证方式
Router-A_config_s1/0#ppp chap hostname RouterA             //设置发送给对方验证的账号
Router-A_config_s1/0#physical-layer speed 64000            //配置 DCE 时钟频率
Router-A_config_s1/0#no shutdown
Router-A_config_s1/0#^Z                                    //按 Ctrl + Z 组合键进入特权模式
```

步骤 2：查看 Router-A 的配置。

```
Router-A#show interface s1/1                               //查看接口状态
Serial1/0 is up, line protocol is down                     //对端没有配置，所以协议处于 DOWN 状态
  Mode=Sync DCE Speed=64000                                //查看 DCE
  DTR=UP,DSR=UP,RTS=UP,CTS=DOWN,DCD=UP
  Interface address is 192.168.1.1/24                      //查看 IP 地址
  MTU 1500 bytes, BW 64 kbit, DLY 2000 usec
  Encapsulation protolol PPP, link check interval is 10 sec   //查看封装协议
  Octets   Received0, Octets Sent 0
  Frames Received 0, Frames Sent 0, Link-check Frames Received0
```

```
        Link-check Frames Sent 89,      LoopBack times 0
        Frames Discarded 0, Unknown Protocols Frames Received 0, Sent failuile 0
            Link-check Timeout 0, Queue Error 0, Link Error 0,
            60 second input rate 0 bits/sec, 0 packets/sec!
            60 second output rate 0 bits/sec, 0 packets/sec!
                0 packets input, 0 bytes, 8 unused_rx, 0 no buffer
                0 input errors, 0 CRC, 0 frame, 0 overrun, 0 ignored, 0 abort
                8 packets output, 192 bytes, 0 unused_tx, 0 underruns
        error:
                0 clock, 0 grace
            PowerQUICC SCC specific errors:
                0 recvallocbmblk fail      0 recv no buffer
            0 transmitter queue full       0 transmitter hwqueue_full
```

步骤3：Router-B 的配置。

Router>enable	//进入特权模式
Router #config	//进入全局配置模式
Router _config#hostname Router-B	//修改机器名
Router-B_config#aaa authentication ppp test local	
//定义一个名为 test，使用本地数据进行验证的 aaa 验证方法	
Router-B_config#usernameRouterA password digitalchina	//设置账号和密码
Router-B_config#interface s1/0	//进入接口模式
Router-B_config_s1/0#ip address 192.168.1.2 255.255.255.0	//配置 IP 地址
Router-B_config_s1/0#encapsulation PPP	//封装 PPP 协议
Router-B_config_s1/0#ppp authentication chap test	//设置验证方式
Router-B_config_s1/0#ppp chap hostname RouterB	//设置发送给对方验证的账号
Router-B_config_s1/0#shutdown	
Router-B_config_s1/0#no shutdown	
Router-B_config_s1/0#^Z	//按 Ctrl + Z 组合键进入特权模式

步骤4：查看 Router-B 的配置。

Router-A#show interface s1/0	//查看接口状态
Serial1/0 is up, line protocol is up	//接口和协议都处于 up 状态
Mode=Sync DTE	//查看 DTE
DTR=UP,DSR=UP,RTS=UP,CTS=DOWN,DCD=UP	
Interface address is 192.168.1.2/24	//查看 IP 地址
MTU 1500 bytes, BW 64 kbit, DLY 2000 usec	
Encapsulation prototol PPP, link check interval is 10 sec	//查看封装协议

```
        Octets    Received0, Octets Sent 0
        Frames Received 0, Frames Sent 0, Link-check Frames Received0
        Link-check Frames Sent 89,      LoopBack times 0
        Frames Discarded 0, Unknown Protocols Frames Received 0, Sent failuile 0
            Link-check Timeout 0, Queue Error 0, Link Error 0,
            60 second input rate 0 bits/sec, 0 packets/sec!
            60 second output rate 0 bits/sec, 0 packets/sec!
                0 packets input, 0 bytes, 8 unused_rx, 0 no buffer
```

```
            0 input errors, 0 CRC, 0 frame, 0 overrun, 0 ignored, 0 abort
            8 packets output, 192 bytes, 0 unused_tx, 0 underruns
error:
            0 clock, 0 grace
            PowerQUICC SCC specific errors:
                0 recvallocbmblk fail         0 recv no buffer
            0 transmitter queue full          0 transmitter hwqueue_full
```

步骤5：测试连通性。

```
Router-A#ping 192.168.1.2
PING 192.168.1.2 (192.168.1.2): 56 data bytes
!!!!!
--- 192.168.1.2 ping statistics ---
5 packets transmitted, 5 packets received, 0% packet loss
round-trip min/avg/max = 20/22/30 ms
```

♂ 小贴士

双方密码一定要一致，发送的账号要和对方账号数据库中的账号对应；不要忘记配置 DCE 的时钟频率。

任务三　实现网络地址转换

♂ 需求分析

某公司向因特网接入运营商申请了专线接入和公网 IP 地址，这意味着公司可以访问因特网了。然而，公司只有一个公网 IP 地址，却有几十台等待联网的计算机。企业内部有对 Internet 提供服务的 Web 服务器，企业内部使用私有地址的主机需要访问 Internet。

♂ 方案设计

当路由器检测到公司内部计算机访问外网的数据包时，会将 IP 数据包头部中的源 IP 地址（即内部保留地址）转换为公司申请的公网 IP 地址（内部全局地址）。经过 IP 地址转换的数据包就可以路由到因特网中了。

所需设备如图 8-3-1 所示。

（1）DCR-2626 路由器 2 台。

（2）PC 2 台。

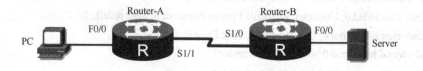

图 8-3-1　实现网络地址转换

设备配置信息，见表 8-3-1。

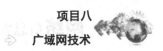

表 8-3-1　设备配置信息

Router-A		Router-B	
F0/0 的 IP 地址	192.168.0.1/24	F0/0 的 IP 地址	192.168.2.1/24
S1/1 (DCE) 的 IP 地址	192.168.1.1/24	S1/0 的 IP 地址	192.168.1.2/24
PC		Server	
IP 地址	192.168.0.3/24	IP 地址	192.168.2.2/24
网关	192.168.0.1	网关	192.168.2.1

知识准备

NAT（Net work Address Translation，网络地址转换）是指将网络地址从一个地址空间转换为另一个地址空间的行为。NAT 将网络划分为内部网络（Inside）和外部网络（Outside）两部分。局域网主机利用 NAT 访问网络时，将局域网内部的本地地址转换为全局地址（互联网合法 IP 地址）后转发数据包。

一般而言，有两种技术可以使用有限的甚至只用一个公网 IP 地址来使大量的主机访问因特网，它们是代理服务器技术和网络地址转换技术。其中，NAT 在大量的末节网络中应用。

网络地址转换是将 IP 包头中的地址进行转换的一种技术，通常用在末节网络的边界网关设备上，用于支持使用私有 IP 地址的内部主机访问外部公网。应用了 NAT 技术，公司不必为每一台内部主机都分配公网地址，即可实现因特网的访问，这大大节约了公网 IP 地址，极大地缓解了 IPv4 地址不足的压力。

因特网接入商大量使用了 NAT 技术。通过网络地址转换，ISP 只需要使用极少的公网 IP 地址，即可为成千上万的用户提供上网服务。

NAT 技术常使用的各类术语如下。

（1）内部本地地址：通常指分配给内部主机使用的私有地址。

（2）内部全局地址：由 ISP 提供的用于本网络使用的公网地址。

（3）外部本地地址：通常指提供给外部网络使用的私有地址。

（4）外部全局地址：外部网络使用的可路由的公网地址。

（5）外部端口：内部网络中运行 NAT 机制的设备，如路由器、防火墙等，用于连接外网的接口。

（6）内部端口：运行 NAT 机制的设备，用于连接内网的接口。

NAT 通常包括以下几种应用类型。

（1）静态 NAT：由网络管理员手工配置，指一对一的私有地址和公网地址之间的转换。

（2）动态 NAT：由设备自动配置，指一段内部地址和一个外部地址池之间的转换，每次转换都是一对一进行的。

（3）静态 NAPT：由网络管理员手工设定，用于将一个内部私有地址和端口号转换为一个公网地址和端口号。静态 NAPT 可以实现一个公网地址的复用。

（4）动态 NAPT：由设备自动设定，利用不同的端口号来将多个私有地址转换为一个公网地址。

（5）NAT 实现 TCP 均衡：使用 NAT 机制创建了一台虚拟主机并提供 TCP 服务，该虚拟主机对应内部多台实际主机，然后对目标地址进行轮询置换，达到负载分流的目的。

动态 NAPT 是最为常见的 NAT 应用。一般情况下，配置 NAPT 应遵循以下步骤。

（1）定义外网端口。

（2）定义内网端口。

（3）定义需要进行地址转换的数据流（使用 IP ACL 来定义）。

（4）定义公网地址池。

（5）建立数据流和公网地址池之间的映射关系。

（6）添加指向外网的静态路由。

任务实现

内部的 PC 需要访问外部的服务器：假设在 Router-A 上做地址转换，将 192.168.0.0/24 转换成 192.168.1.10～192.168.1.20 之间的地址，并且做端口的地址复用。

步骤 1：按相关任务和表 8-3-1 将接口地址和 PC 地址配置好，并做连通性测试。

步骤 2：配置 Router-A 的 NAT。

```
Router-A#conf
Router-A_config#ip access-list standard 1                    //定义标准访问控制列表
Router-A_config_std_nacl#permit 192.168.0.0 255.255.255.0
//定义允许转换的源地址范围
Router-A_config_std_nacl#exit
Router-A_config#ip nat pool overld 192.168.1.10 192.168.1.20 255.255.255.0
//定义名为 overld 的转换地址池
Router-A_config#ip nat inside source list 1 pool overld overload
//将 ACL 允许的源地址转换成 overld 中的地址，并且做 PAT 的地址复用
Router-A_config#int f0/0
Router-A_config_f0/0#ip nat inside                           //定义 F0/0 为内部接口
Router-A_config_f0/0#int s1/1
Router-A_config_s1/1#ip nat outside                          //定义 S1/1 为外部接口
Router-A_config_s1/1#exit
Router-A_config#ip route 0.0.0.0 0.0.0.0 192.168.1.2 //配置路由器 A 的默认路由
```

步骤 3：查看 Router-B 的路由表。

```
Router-B#sh ip route
Codes: C - connected, S - static, R - RIP, B - BGP, BC - BGP connected
       D - DEIGRP, DEX - external DEIGRP, O - OSPF, OIA - OSPF inter area
       ON1 - OSPF NSSA external type 1, ON2 - OSPF NSSA external type 2
       OE1 - OSPF external type 1, OE2 - OSPF external type 2
       DHCP - DHCP type
VRF ID: 0
C      192.168.1.0/24        is directly connected, Serial1/0
C      192.168.2.0/24        is directly connected, FastEthernet0/0
//注意，并没有到 192.168.0.0 的路由
```

步骤 4：测试网络是否可以通信，如图 8-3-2 所示。

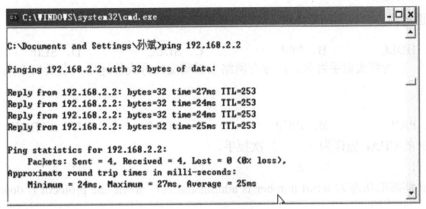

图 8-3-2 测试连通性

步骤 5：查看地址转换表。

```
Router-A#sh ip nat translatios
Pro.DirInside local          Inside global        Outside local        Outside global
ICMP   OUT 192.168.0.3:512    192.168.1.10:12512   192.168.1.2:12512    192.168.1.2:12512
```

认证考核

一、选择题

1. 在路由器上进行广域网连接时，必须设置的参数是（　　）。
   A. 在 DTE 端设置 clock rate
   B. 在 DCE 端设置 clock rate
   C. 在路由器上配置远程登录
   D. 添加静态路由

2. 下列关于 HDLC 的说法，错误的是（　　）。
   A. HDLC 运行于同步串行线路
   B. HDLC 协议的单一链路只能承载单一的网络层协议
   C. HDLC 是面向字符的链路层协议，其传输的数据必须是规定字符集中的字符
   D. HDLC 是面向比特的链路层协议，其传输的数据必须是规定字符集中的字符

3. HDLC 是一种面向（　　）的链路层协议。
   A. 字符　　　　　B. 比特　　　　　C. 信元　　　　　D. 数据包

4. 下列所述的协议中，不是广域网协议的是（　　）。
   A. PPP　　　　　B. X.25　　　　　C. HDLC　　　　　D. RIP

5. 下列关于 PPP 协议的说法，正确的是（　　）。
   A. PPP 协议是一种 NCP 协议
   B. PPP 协议与 HDLC 同属于广域网协议
   C. PPP 协议只能工作在同步串行链路上
   D. PPP 协议是三层协议

6. 下面的封装协议使用 CHAP 或者 PAP 验证方式的是（　　）。

    A．HDLC   B．PPP   C．SDLC   D．SLIP

7. （　　）为两次握手协议，通过在网络上以明文的方式传递用户名及口令来对用户进行验证。

    A．PAP   B．IPCP   C．CHAP   D．RADIUS

8. PPP 的 CHAP 验证为（　　）次握手。

    A．1   B．2   C．3   D．4

9. PPP 链路的状态为 serial number is administratively down,line protocol is down 时，说明（　　）。

    A．物理链路有问题   B．接口被管理员停用了
    C．参数配置不正确   D．没有大问题，重启路由器即可

二、实训题

在 R2 的 F0/0 接口上添加 ACL，禁止来自网段 192.168.0.0/24 的数据包。在 R1 路由器上配置 NAPT，使左侧内部主机 PC1 可以通过公网访问右侧的 Web Server。

实验拓扑如图 8-3-3 所示。

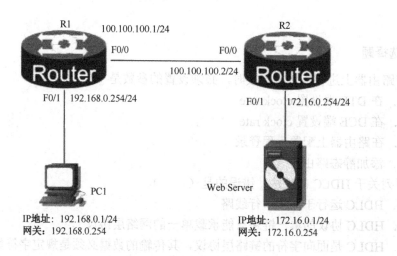

图 8-3-3　NAPT 训练用图

项目九 网络安全与 VPN 实现

♂ 教学背景

VPN（Virtual Private Network，虚拟专用网）被定义为通过一个公用网络（通常是因特网）建立一个临时的、安全的连接，是一条穿过混乱的公用网络的安全、稳定的隧道。虚拟专用网是对企业内部网的扩展。虚拟专用网可以帮助远程用户、公司分支机构、商业伙伴及供应商同公司的内部网建立可信的安全连接，并保证数据的安全传输。虚拟专用网可用于不断增长的移动用户的全球互联网接入，以实现安全连接；可用于实现企业网站之间安全通信的虚拟专用线路；可用于经济有效地连接到商业伙伴和用户。

防火墙指的是一个由软件和硬件设备组合而成、在内部网和外部网之间、专用网与公共网之间的界面上构造的保护屏障，是一种获取安全性方法的形象说法，它是一种计算机硬件和软件的结合，使 Internet 与 Intranet 之间建立起一个安全网关（Security Gateway），从而保护内部网免受非法用户的侵入。

任务一 路由器使用 PPTP 实现 VPDN

♂ 需求分析

某公司内部网络将一台路由器作为公司局域网网关，并使用 PPTP 提供 VPDN 拨入服务。现要在路由器上做适当配置，实现公司外部主机与公司内部网络中的主机之间的安全通信。

♂ 方案设计

PPTP（Point to Point Tunneling Protocol，点对点隧道协议）是一种支持多协议虚拟专用网的网络技术。通过该协议，远程用户能够通过 Microsoft Windows NT 工作站、Windows 95 和 Windows 98 操作系统及其他装有点对点协议的系统安全访问公司网络，并能拨号连入本地 ISP，通过 Internet 安全连接到公司网络。本任务以 PC1 为远程客户端创建到 LNS 的 PPTP 隧道，并访问公司内部的服务器 PC2。

所需设备如图 9-1-1 所示。

（1）PC 2 台。
（2）路由器 1 台。

（3）交叉双绞线 2 条。
（4）路由器配置线缆 1 条。

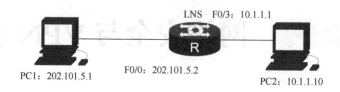

图 9-1-1　路由器使用 PPTP 实现 VPDN

知识准备

PPTP 的默认端口号为 1723。该协议是在 PPP 的基础上开发的一种新的增强型安全协议，支持多协议虚拟专用网，可以通过密码验证协议、可扩展认证协议（EAP）等方法增强安全性；可以使远程用户通过拨入 ISP、直接连接 Internet 或其他网络安全地访问企业网。

任务实现

步骤 1：配置用户和地址池。

LNS_config#username aaa password 0 123
//创建一个账户，用户名为 aaa，密码为 123
LNS_config#aaa authentication ppp pptp local
//创建一个 aaa 列表，表名为 pptp，此列表用于 PPP 验证过程，并进行本地（local）验证
LNS_config#ip local pool pptp 10.1.4.2 10
//创建 VPN 拨入的地址池，命名为 pptp，范围为 10.1.4.2～10.1.4.11

步骤 2：配置路由器接口地址。

LNS_config#interface fastEthernet 0/0
LNS_config_f0/0#ip add 202.101.5.2 255.255.255.0
LNS_config_f0/0#exit
LNS_config#interface fastEthernet 0/3
LNS_config_f0/3#ip add 10.1.1.1 255.255.255.0

步骤 3：配置虚拟模板。

LNS_config#interface virtual-template 0   //创建虚模板 0
LNS_config_vt0#ppp authenentication pap pptp   /*加密方式为 PAP，调用 PPTP 的 aaa 列表进行本地验证*/
LNS_config_vt0#ip add 10.1.4.1 255.255.255.0   //配置隧道网关地址
LNS_config_vt0#ppp pap sent-username aaa password 123
//设置隧道端口 PPP 验证时发送的用户名与口令（可选配置）
LNS_config_vt0#peer default ip address pool pptp
//设置拨入用户分配的地址范围，调用 PPTP 地址池

步骤 4：配置 VPN 相关参数。

LNS_config#vpdn enable
LNS_config#vpdn-group 0   //创建一个 VPDN 组，组号为 0
LNS_config_vpdn_0#accept-dialin   //允许拨号
LNS_config_vpdn_0#protocol pptp   //协议为 PPTP
LNS_config_vpdn_0#port virtual-template 0   //引用虚模板 0

步骤 5：配置默认路由。
```
LNS_config#ip route default 202.101.5.1        //设定默认路由
```
步骤 6：设置 PC1 的 IP 地址为 202.101.5.1，不必配置网关。

在 PC1 的"网络和拨号连接"中新建连接，"网络连接类型"选择"通过 Internet 连接到专用网络"，"公用网络"选择"不拨初始连接"，目标地址为 202.101.5.2，命名此连接为 Dialer。在 Dialer 的属性中，指定 VPDN 服务器的类型为 PPTP，并在安全设置中定义使用 PAP 认证及可选加密。在单击"呼叫"按钮时，输入路由器所配置的用户名 aaa 和密码 123。配置完成，PC1 拨入路由器后，即可访问内部网的服务器。

♂ 注意事项和排错

注意，PC1 中不配置默认网关。拨叫成功后，在 LNS 控制面板中会出现如下信息。

```
LNS_config#Jan   1 00:38:49 Line on Interface Virtual-access0, changed state to up
Jan   1 00:38:51 Line protocol on Interface Virtual-access0, changed state to up
/*注意，原配置的模板 virtual-templete 0 已经派生出了一个具体的接口，即 Virtual-access0*/
LNS#sh interface virtual-access 0           //显示当前已经建立的虚拟隧道接口情况
Virtual-access0 is up, line protocol is up
    Hardware is Virtual access interface
    MTU 1400 bytes, BW 100000 kbit, DLY 10000 usec
    Interface address is 10.1.4.1/24
    Encapsulation PPP, loopback not set
Keepalive set(10 sec)
    LCP   Opened
    PAP   Opened,  Message: 'none'
    IPCP Opened
        local IP address: 10.1.4.1    remote IP address: 10.1.4.2
LNS#sh pptp tunnel                        //显示当前 PPTP 的隧道情况
PPTP Tunnel Information Total tunnels 1 sessions 1
Socket    TunlID    Remote Name       State             Sessions Remote Address
17        1                           Established       1        202.101.5.1
LNS#sh pptp session                      //显示当前 PPTP 的会话状态
PPTP Session Information Total tunnels 1 sessions 1
LocID PeerID TunlID Intf      State
1     31729  1      va0       Established
```

为使远程 VPDN 用户能够访问公司内部的服务器，这些服务器必须配置可以到达远程拨入用户的路由。一般情况下，只要将这些服务器的默认网关设置为路由器的内部网关地址即可。

任务二 使用 L2TP 连接企业总部与分支机构

♂ 需求分析

某公司内部网络通过一台路由器连接到公司外的另一台路由器上，现要在路由器上做适当配置，实现公司总部的主机与公司分部主机之间的安全通信。

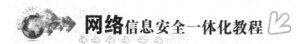

方案设计

本任务通过在 Router1 和 Router2 上配置 VPN，由 Router1 作为 LNS（L2TP Network Server，L2TP 网络服务器），Router2 作为 LAC（L2TP Access Concentrator，L2TP 访问集中器），通过 L2TP 的 VPN 方式保护总部与分部之间的数据通信。

所需设备如图 9-2-1 所示。

（1）DCR-2626 路由器 2 台。
（2）交叉双绞线 3 条。
（3）PC 2 台。
（4）路由器控制线缆 1 条。

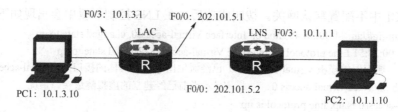

图 9-2-1　使用 L2TP 连接企业总部与分支机构

知识准备

L2TP 是一种工业标准的 Internet 隧道协议，功能大致和 PPTP 类似，如同样可以对网络数据流进行加密。但它们也有不同之处，如 PPTP 要求网络为 IP 网络，L2TP 要求面向数据包的点对点连接；PPTP 使用单一隧道，L2TP 使用多隧道；L2TP 提供包头压缩、隧道验证，而 PPTP 不支持。

PPTP 和 L2TP 都使用 PPP 对数据进行封装，然后添加附加包头用于数据在互联网中的传输。尽管两个协议非常相似，但是仍存在以下不同。

（1）PPTP 要求互联网为 IP 网络。L2TP 只要求隧道媒介提供面向数据包的点对点的连接。L2TP 可以在 IP（使用 UDP）、帧中继永久虚拟电路（PVCs）、X.25 虚拟电路（VCs）或 ATM VCs 网络中使用。

（2）PPTP 只能在两端点间建立单一隧道。L2TP 支持在两端点间使用多隧道。使用 L2TP，用户可以针对不同的服务质量创建不同的隧道。

（3）L2TP 可以提供包头压缩，当压缩包头时，系统开销占用 4 个字节，而 PPTP 协议下要占用 6 个字节。

（4）L2TP 可以提供隧道验证，而 PPTP 不支持隧道验证。但是当 L2TP 或 PPTP 与 IPSec 共同使用时，可以由 IPSec 提供隧道验证，不需要在第二层协议上验证隧道。

（5）L2TP 访问集中器是一种附属在网络上的具有 PPP 端系统和 L2TP v2 协议处理能力的设备，它一般是一个网络接入服务器软件，在远程客户端完成网络接入服务的功能。

（6）L2TP 网络服务器是用于处理 L2TP 服务器端各种情况的软件。L2TP 支持的协议有 IP、IPX 和 NetBEUI。

项目九 网络安全与 VPN 实现

♂ 任务实现

步骤 1：配置用户和地址池。

```
LNS_config#username aaa password 0 123
//创建一个账户，用户名为 aaa，密码为 123
LNS_config#aaa authentication ppp l2tp local
/*创建一个 aaa 列表，表名为 l2tp，此列表用于 PPP 验证过程，并进行本地（local）验证*/
LNS_config#ip local pool l2tp 10.1.4.2 10
//创建 VPN 拨入的地址池，命名为 l2tp，范围为 10.1.4.2～10.1.4.11
LAC_config#username aaa pass 123
//创建一个账户，用户名为 aaa，密码为 123
LAC_config#aaa authentication ppp l2tp local
/*创建一个 aaa 列表，表名为 l2tp，此列表用于 PPP 验证过程，并进行本地（local）验证*/
```

步骤 2：配置路由器接口地址。

```
LNS_config#interface fastEthernet 0/0
LNS_config_f0/0#ip add 202.101.5.2 255.255.255.0
LNS_config_f0/0#exit
LNS_config#interface fastEthernet 0/3
LNS_config_f0/3#ip add 10.1.1.1 255.255.255.0
LAC_config#int f 0/0
LAC_config_f0/0#ip add 202.101.5.1 255.255.255.0
LAC_config_f0/0#exit
LAC_config#interface fastEthernet 0/3
LAC_config_f0/3#ip add 10.1.3.1 255.255.255.0
```

步骤 3：配置虚拟模板。

```
LNS_config#interface virtual-template 0              //创建虚模板 0
LNS_config_vt0#ppp authentication pap l2tp
//加密方式为 PAP，调用 l2tp 的 aaa 列表进行本地验证
LNS_config_vt0#ip add 10.1.4.1 255.255.255.0         //配置隧道网关地址
LNS_config_vt0#peer default ip address pool l2tp
//设置拨入用户分配的地址范围，调用 l2tp 地址池
LNS_config_vt0#ppp pap sent-username aaa password 123
//设置隧道端口 PPP 验证时发送的用户名和口令（可选配置）
LAC_config#interface virtual-tunnel 0                //创建虚模板 0
LAC_config_vn0#ppp authentication pap l2tp
//加密方式为 PAP，调用 l2tp 的 aaa 列表进行本地验证
LAC_config_vn0#ip address negotiated                 //配置隧道地址自动获取
LAC_config_vn0#ppp pap sent-username aaa password 123
//设置隧道端口 PPP 验证时发送的用户名和口令
```

步骤 4：配置 VPN 相关参数。

```
LNS_config#vpdn enable
LNS_config#vpdn-group 0                              //创建一个 VPDN 组，组号为 0
LNS_config_vpdn_0#accept-dialin                      //允许拨号
LNS_config_vpdn_0#protocol l2tp                      //协议为 L2TP
```

153 | PAGE

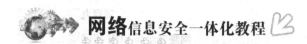

```
LNS_config_vpdn_0#port virtual-template 0        //引用虚模板 0
LAC_config#vpdn enable
LAC_config#vpdn-group 0                          //创建一个 VPDN 组，组号为 0
LAC_config_vpdn_0#request-dialin                 //请求拨号
LAC_config_vpdn_0#protocol l2tp                  //协议为 L2TP
LAC_config_vpdn_0#initiate-to ip 202.101.5.2 priority 0
//向 202.101.5.2 发起 L2TP 连接请求，配置当前优先级为最高
LAC_config_vpdn_0#port virtual-tunnel 0          //引用虚模板 0
```

步骤 5：配置去往分支机构的内网路由。

```
LNS_config#ip route 10.1.3.0 255.255.255.0 virtual-access 0
//设置总部去往 10.1.3.0 分支网段的出口为隧道接口
LAC_config#ip route 10.1.0.0 255.255.0.0 virtual-tunnel 0
//设置分支去往所有公司其他内网网段的出口为此隧道接口
```

注意事项和排错

　　Server 端和 Client 端的参数应该一致。如果在路由器两端都配置了通道认证，则必须在两边通道上配置一致的密码，认证才能成功。本任务中 PC1 不配置默认网关，只添加如下路由：

```
C:\>Route add 202.101.5.0 mask 255.255.255.0 10.1.3.1
LNS#sh interface virtual-access 0        //查看当前建立的虚拟隧道情况
Virtual-access0 is up, line protocol is up
    Hardware is Virtual access interface
    MTU 1500 bytes, BW 100000 kbit, DLY 10000 usec
    Interface address is 10.1.4.1/24
    Encapsulation PPP, loopback not set
  Keepalive set(10 sec)
    LCP    Opened
    PAP    Opened,   Message: 'Welcome to Digitalchina Router'
    IPCP Opened
        local IP address: 10.1.4.1    remote IP address: 10.1.4.2
LNS#
LAC#sh int virtual-tunnel 0              //查看虚拟隧道情况
Virtual-tunnel0 is up, line protocol is up
    Hardware is Unknown device
    MTU 1500 bytes, BW 100000 kbit, DLY 10000 usec
    Internet address is 10.1.4.2/32
    Encapsulation PPP, loopback not set
  Keepalive set(10 sec)
    LCP    Opened
    PAP    Opened,   Message: 'Welcome to Digitalchina Router'
    IPCP Opened
        local IP address: 10.1.4.2    remote IP address: 10.1.4.1
LAC#
LNS#sh l2tp tunnel                       //查看 L2TP 隧道参数
L2TP Tunnel Information:
```

```
Total tunnels 1 sessions 1
Local_ID  Remote_ID  State  Sessions  Remote_Name  UDP_Port  Remote_Address
1         1          Est    1         vpdn-group   1701      202.101.5.1
LNS#
LAC#sh l2tp tunnel                                //查看 L2TP 隧道参数
L2TP Tunnel Information:
Total tunnels 1 sessions 1
Local_ID  Remote_ID  State  Sessions  Remote_Name  UDP_Port  Remote_Address
1         1          Est    1         vpdn-group   1701      202.101.5.2
```

任务三　防火墙初级管理

需求分析

某公司规模不断扩大，为了提高网络的安全性，公司经理让网络管理员购买一种可以满足网络安全需求的设备，网络管理员认为购买 DCFW-1800S-H-V2 防火墙可以满足公司的网络安全需求。

方案设计

防火墙是当今使用最为广泛的安全设备，防火墙历经几代发展，现今为非常成熟的硬件体系结构，具有专门的 Console 口、专门的区域接口，串行部署于 TCP/IP 网络中，将网络划分为内、外、服务器区 3 个区域，对各区域实施安全策略以保护重要网络。

V2 防火墙可以使用 Telnet、SSH、WebUI 等方式进行管理，使用者可以很方便地使用几种方式进行管理。本任务使用 DCFW-1800S-H-V2 防火墙。

所需设备如图 9-3-1 所示。

（1）防火墙 DCFW-1800S-H-V2 1 台。
（2）Console 线 1 条。
（3）网线 2 条。
（4）PC 1 台。

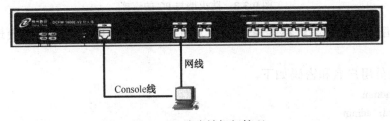

图 9-3-1　防火墙初级管理

任务要求：熟悉防火墙各接口及其连接方法；熟练使用各种线缆实现防火墙与主机、交换机的连通；使用控制台连接防火墙并进行初始配置；掌握防火墙管理环境的搭建和配置方法，熟练使用各种管理方式管理防火墙。

知识准备

V2 防火墙默认的管理员名称是 admin，可以对其进行修改，但不能删除这个账号。
增加一个管理员的命令如下。

```
DCFW-1800(config)#admin user user-name
```

执行该命令后，系统创建指定名称的管理员，并且进入管理员配置模式；如果指定的管理员名称已经存在，则直接进入管理员配置模式。

管理员特权为管理员登录设备后拥有的权限。DCF OS 允许的权限有 RX 和 RXW。
在管理员配置模式下，输入以下命令配置管理员的特权。

```
DCFW-1800(config-admin)#privilege {RX | RXW}
```

在管理员配置模式下，输入以下命令配置管理员的密码。

```
DCFW-1800(config-admin)#password password
```

任务实现

1. 初步认识防火墙

步骤 1：认识防火墙各接口，理解防火墙各接口的作用，并学会使用线缆连接防火墙和交换机、主机，如图 9-3-2 所示。

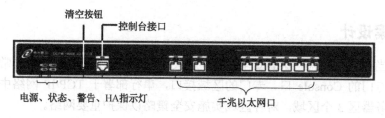

图 9-3-2　初步认识防火墙

步骤 2：使用控制线缆对防火墙与 PC 的串行接口进行连接，如图 9-3-3 所示。

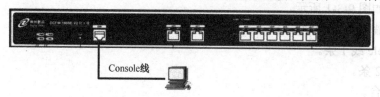

图 9-3-3　防火墙与 PC 的连接

步骤 3：配置 PC 的超级终端属性，进入防火墙命令行模式，登录防火墙并熟悉各配置模式。

默认管理员用户名和密码如下。

```
login: admin
password: admin
```

输入如上信息，可进入防火墙的执行模式，该模式的提示符为

```
DCFW-1800#
```

在执行模式下，输入 configure 命令，可进入全局配置模式，即

```
DCFW-1800(config)#
```

V2 系列防火墙的不同模块功能需要在其对应的命令行子模块的模式下进行配置。在全局配置模式下输入特定的命令可以进入相应的子模块配置模式。例如，运行 interface ethernet0/0 命令可进入 ethernet0/0 接口配置模式，此时的提示符变更为

```
DCFW-1800(config-if-eth0/0)#
```

步骤 4：通过 PC 测试与防火墙的连通性。

使用交叉双绞线连接防火墙和 PC，此时防火墙的 LAN-link 灯亮起，表明网络的物理连接已经建立。观察指示灯状态为闪烁，表明有数据在尝试传输。

此时，打开 PC 的连接状态，发现只有数据发送，没有接收到数据，这是因为防火墙的端口默认状态下会禁止向未经验证和配置的设备发送数据，以保证数据的安全。

2. 搭建 Telnet 和 SSH 管理环境

步骤 1：运行 manage telnet 命令启用被连接接口的 Telnet 管理功能。

```
Hostname#configure
DCFW-1800(config)#interface Ethernet 0/0
DCFW-1800(config-if-eth0/0)#manage telnet
```

步骤 2：运行 manage ssh 命令启用 SSH 管理功能。

```
DCFW-1800(config-if-eth0/0)#manage ssh
```

步骤 3：配置 PC 的 IP 地址为 192.168.1.*，从 PC 上尝试与防火墙的 Telnet 连接。注意，用户名和密码是默认管理员用户名和密码，如图 9-3-4 和图 9-3-5 所示。

图 9-3-4　Telnet 防火墙界面

图 9-3-5　输入用户名和密码后连接成功

3. 搭建 WebUI 管理环境

步骤 1：初次使用防火墙时，用户可以通过 E0/0 接口访问防火墙的 WebUI 页面。在浏览器地址栏中输入"https://192.168.1.1"并按 Enter 键，系统 WebUI 的登录界面如图 9-3-6 所示。

图 9-3-6　防火墙 Web 登录界面

步骤2：登录成功后的主界面如图9-3-7所示。

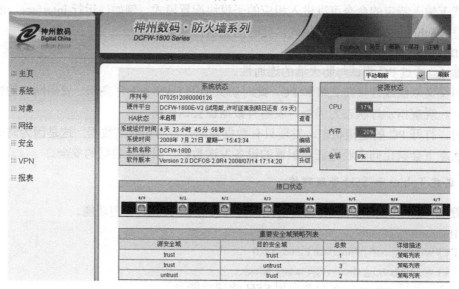

图9-3-7　登录成功后的主界面

任务四　防火墙典型环境安全策略实施

♂ 需求分析

某公司为了保证内部网络的安全性，购买了防火墙，考虑到公网地址有限，从节约成本的角度讲，不可能每台PC都配置公网地址访问外网。网络管理员想使用防火墙的混合模式来通过少量公网IP地址满足多数私网IP地址的联网，以缓解IP地址不足和节约成本。

♂ 方案设计

混合模式相当于防火墙既工作于路由模式，又工作于透明模式。在实际应用环境中，此类防火墙应用也比较广泛。混合模式分为两种：ISP分配外网地址，内网为私有地址，服务器区域和内部地址为同一网段，这样，内部区域和服务器区域为透明，内部区域和外网区域为路由，服务器区域和外部区域也为路由；ISP分配外网地址，内网为私有地址，服务器区域使用ISP分配的公网地址，服务器区域和外网为透明，内部区域和外部区域为路由。在下面的应用环境中，使用的是第二种混合模式。

所需设备如图9-4-1所示。

（1）防火墙设备1台。
（2）Console线1条。
（3）网线2条。
（4）PC 1台。

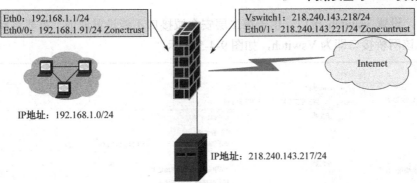

图 9-4-1　公司网络拓扑结构

任务实现

步骤 1：设置内网地址，设置 Eth0 为内网口，IP 地址为 192.168.1.1/24，如图 9-4-2 所示。

图 9-4-2　设置内网口地址

步骤 2：设置外网口，Eth6 口连接外网，将 Eth6 口设置成第二层安全域 l2-untrust，如图 9-4-3 所示。

图 9-4-3　设置外网口

步骤 3：设置服务器接口，将 Eth7 口设置成 l2-dmz 安全域，连接服务器，如图 9-4-4 所示。

图 9-4-4　设置服务器接口

步骤 4：设置 vswitch 接口。由于第二层安全域接口不能设置地址，需要将地址设置在网桥接口上，该网桥接口即为 Vswitch，如图 9-4-5 所示。

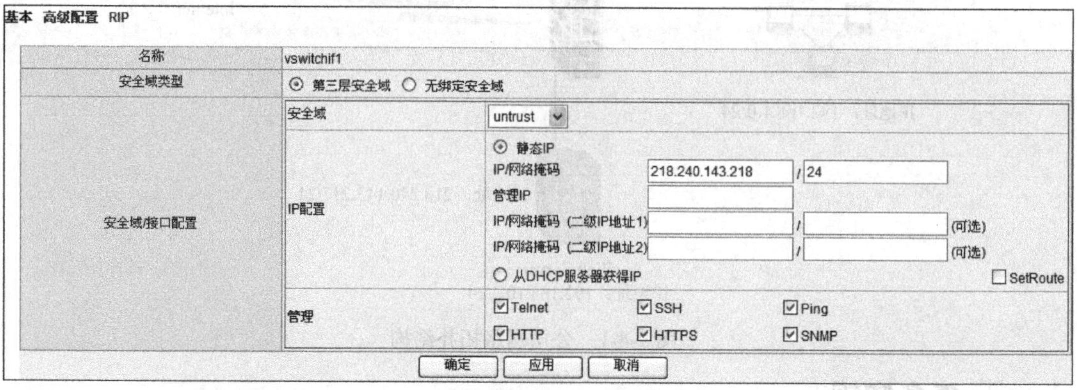

图 9-4-5 设置 Vswitch 接口

步骤 5：设置 SNAT 策略。针对内网所有地址，在防火墙上设置源 NAT，内网 PC 在访问外网时，凡是从 vswitch 接口出去的数据包都做地址转换，转换地址为 Vswitch 接口地址，如图 9-4-6 所示。

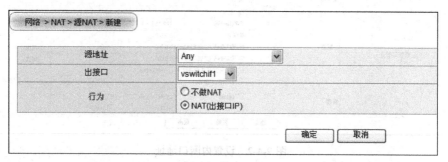

图 9-4-6 设置 SNAT 策略

步骤 6：添加路由。要创建一条到外网的默认路由，如果内网有三层交换机，则需要创建到内网的回指路由，如图 9-4-7 所示。

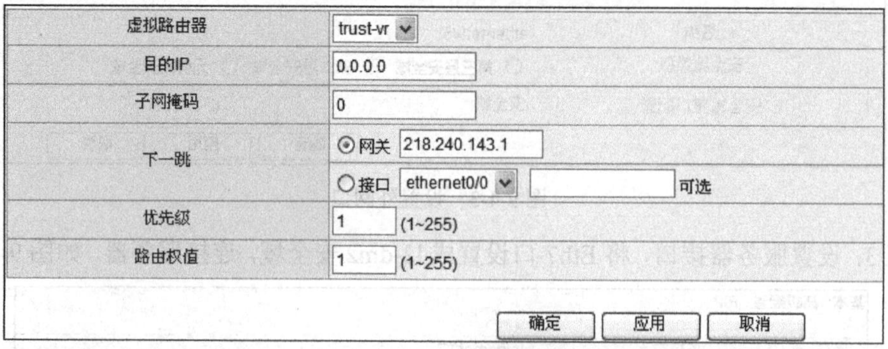

图 9-4-7 添加路由

步骤 7：设置地址簿。在放行安全策略时，需要选择相应的地址和服务进行放行，所以这里先要创建服务器的地址簿。在创建地址簿时，如果创建的服务器属于单个 IP，使用 IP 成员方式，则网络掩码一定要写 32 位，如图 9-4-8 所示。

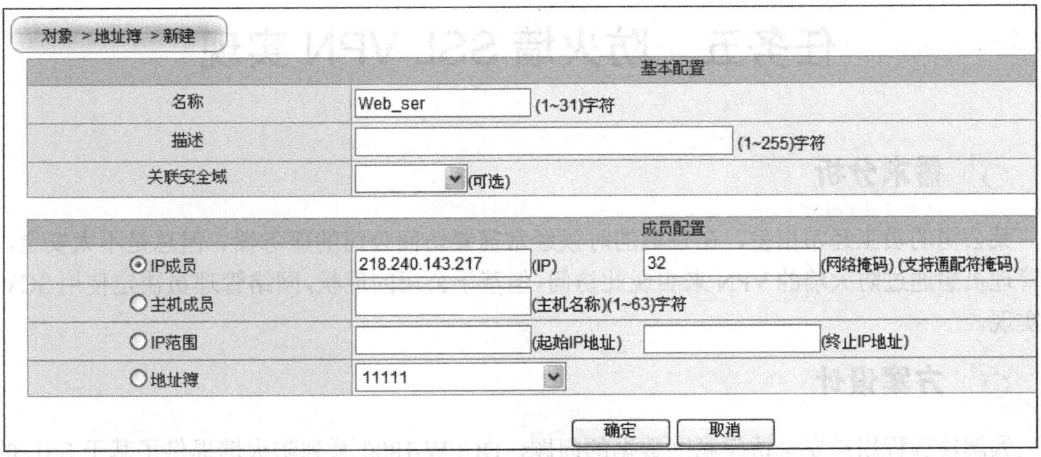

图 9-4-8 设置地址簿

步骤 8：设置放行策略。放行策略时，首先要保证内网能够访问外网。应该放行内网口所属安全域到 vswitch 接口所属安全域的安全策略，即应该是从 trust 到 untrust，如图 9-4-9 所示。

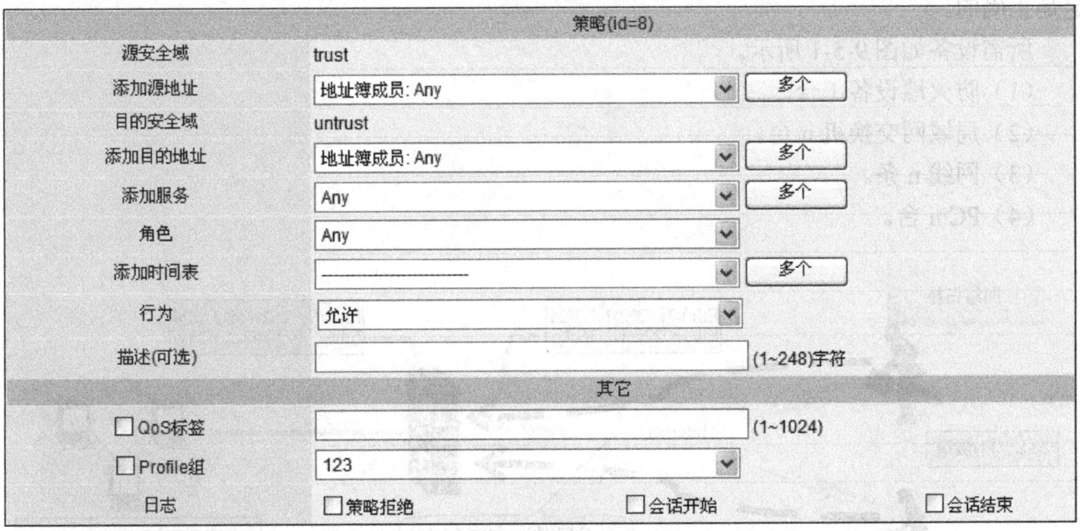

图 9-4-9 设置放行策略

步骤 9：要保证外网能够访问 Web_server，该服务器的网关地址应设置为 ISP 网关，即 218.240.143.1。若需要放行第二层安全之前的安全策略，则应该放行 l2-untrust 到 l2-dmz 策略。

小贴士

如果服务器的网关并未设置成 218.240.143.1，而设置成 Vswitch 接口地址，此时安全策略如何设置才能使外网访问 Web 服务器？

按上述任务配置完成后，可以在外网访问服务器，那么在服务器上是否可以访问外网呢？如果无法访问，则是什么原因，如何才能实现？

任务五 防火墙 SSL VPN 实现

需求分析

某公司的员工经常出差,在外地的时候经常需要访问公司的服务器,但总是不太安全,网络管理员想通过防火墙的 VPN 来实现此访问,但基于公司的现状,网络管理员决定使用 SCVPN 来实现。

方案设计

为解决远程用户安全访问私网数据的问题,DCFW-1800 系列防火墙提供了基于 SSL 的远程登录解决方案——Secure Connect VPN,简称为 SCVPN。SCVPN 可以通过简单易用的方法实现信息的远程连通。防火墙的 SCVPN 功能包含设备端和客户端两部分,配置了 SCVPN 功能的防火墙作为设备端。外网用户通过 Internet 使用 SSL VPN 接入内网。所以网络管理员的决定是正确的。

所需设备如图 9-5-1 所示。
(1) 防火墙设备 1 台。
(2) 局域网交换机 n 台。
(3) 网线 n 条。
(4) PC n 台。

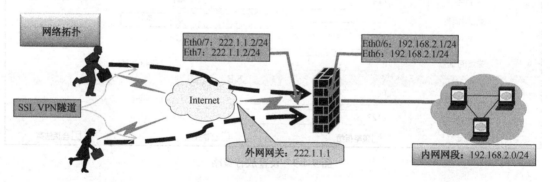

图 9-5-1 防火墙 SSL VPN 拓扑结构

知识准备

SSL VPN 指的是基于安全套接层协议(Security Socket Layer,SSL)建立远程安全访问通道的 VPN 技术。它是近年来兴起的 VPN 技术,其应用随着 Web 的普及和电子商务、远程办公的兴起而发展迅速。

SSL VPN 由于其强大的功能和实施的方便性而应用得越来越广泛,市场上的 SSL VPN 品牌也越来越多,如何选择适合自己的产品是需要用户仔细考虑的一个问题。

SSL VPN 的发展迎合了用户对低成本、高性价比远程访问的需求。现在,它已经广泛应用于各行各业。选购 SSL VPN 时,用户要根据自身特点和不同的业务模式,选择适合自己的 SSL VPN 产品,再次强调,VPN 是正在发展的技术,更新换代可能会比较快,因此用户在选购时可以少考虑一些扩展性，多注重产品的实用性。毕竟，只有适合自己的，才是最理想的选择。

SSL 协议主要是由 SSL 记录协议和握手协议组成的，它们共同为应用访问连接提供认证、加密和防篡改功能。SSL 握手协议相对于 IPSec 协议体系中的 IKE（互联网密钥交换）协议，主要用于服务器和客户之间的相互认证，协商加密算法和 MAC（Message Authentication Code，消息认证码）算法，用于生成在 SSL 记录协议中使用的加密和认证密钥。SSL 记录协议用于为各种应用协议提供基本的安全服务，类似于 IPSec 的传输模式，应用程序消息参照 MTU（最大传输单元）被分割成可治理的数据块（可进行数据压缩处理），并产生一个 MAC 信息，加密后插入新的报头，最后在 TCP 中加以传输；接收端将收到的数据解密，做身份验证（MAC 认证）、解压缩、重组数据报并交给应用协议进行处理。实际上，就是在应用层和传输层之间加入了一个数据处理层，并和传统的网络套接字模型在同一层次，这也是安全套接层的由来。

任务实现

步骤 1：配置 SCVPN 地址池，在 VPN/SCVPN/地址池中新建一个名为"scvpn_pool1"的地址池，如图 9-5-2 所示。

图 9-5-2　配置 SCVPN 地址池

步骤 2：配置 SCVPN 实例，在 VPN/SCVPN/配置中，创建一个 SCVPN 实例，定义 SCVPN 接入使用的各种参数，如图 9-5-3 所示。

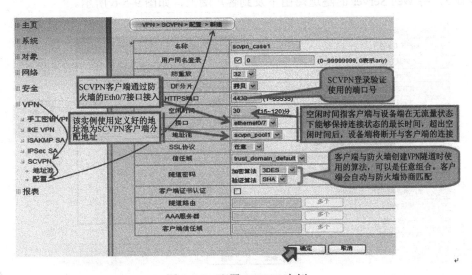

图 9-5-3　配置 SCVPN 实例

步骤 3：创建完 SCVPN 实例并编辑完各种参数后，还需要对该实例重新进行编辑。单击已编辑好的实例右侧的"修改"按钮，如图 9-5-4 所示。

图 9-5-4 查看实例配置参数

步骤 4：添加隧道路由条目。在客户端与防火墙的 SCVPN 创建成功后会下发到客户端的路由表中。添加的网段就是客户端要通过 VPN 隧道访问的位于防火墙内网的网段。需要注意的是，此处添加的路由条目的"度量"值比客户端上默认路由的度量值小。度量值越小的路由条目优先级越高，如图 9-5-5 所示。

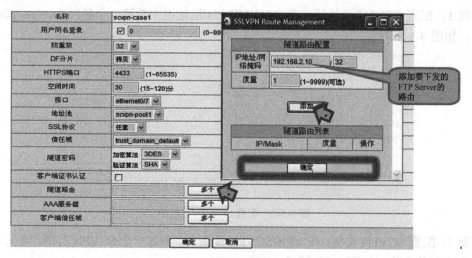

图 9-5-5 添加隧道路由条目

步骤 5：将 Web Server 的隧道路由下发到客户端上，如图 9-5-6 所示。

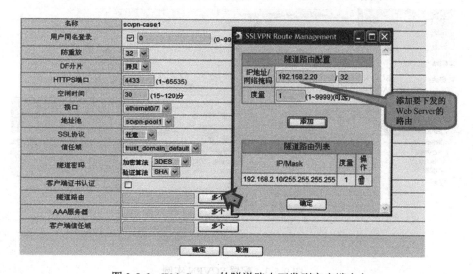

图 9-5-6 Web Server 的隧道路由下发到客户端上

步骤 6：添加的 AAA 服务器用来验证客户端登录的用户名、密码。目前，防火墙支持的验证方式有 4 种：防火墙本地验证、RADIUS 验证、Active-Directory 验证和 LDAP 验证，如图 9-5-7 所示。

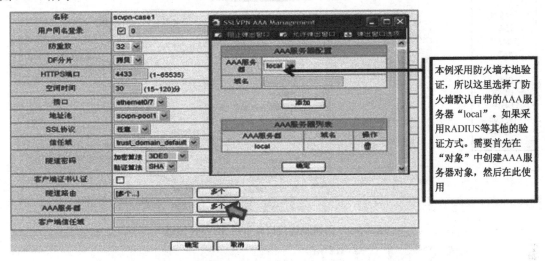

图 9-5-7　添加的 AAA 服务器

步骤 7：创建 SCVPN 所属安全域，在网络/安全域中为创建的 SCVPN 新建一个安全域，安全域类型为"第三层安全域"，如图 9-5-8 所示。

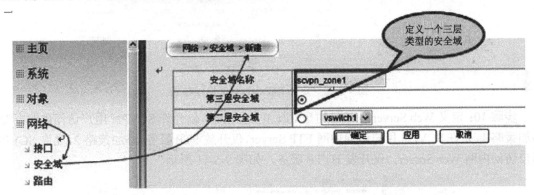

图 9-5-8　创建 SCVPN 所属安全域

步骤 8：创建隧道接口并引用 SCVPN 隧道。为了使 SCVPN 客户端与防火墙上其他接口所属区域之间正常路由转发，需要为它们配置一个网关接口，这在防火墙上通过创建一个隧道接口，并将创建好的 SCVPN 实例绑定到该接口上来实现，如图 9-5-9 所示。

步骤 9：创建安全策略。在放行安全策略前，首先要创建地址簿和服务簿。为创建 SCVPN 客户端访问内网 Server 的安全策略，首先要将策略中引用的对象定义好，如图 9-5-10 所示。

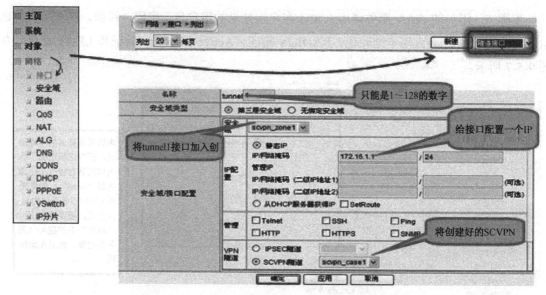

图 9-5-9 创建隧道接口并引用 SCVPN 隧道

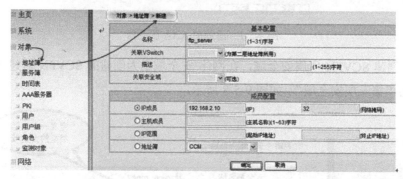

图 9-5-10 定义 FTP Server 的地址对象

步骤 10：定义 Web Server 的地址对象。添加安全策略，以允许 SCVPN 用户访问内网资源：添加策略 1，允许 SCVPN 用户访问内网 FTP Server，仅开放 FTP 服务；添加策略 2，允许 SCVPN 用户访问内网 Web Server，仅开发 HTTP 服务，如图 9-5-11 所示。

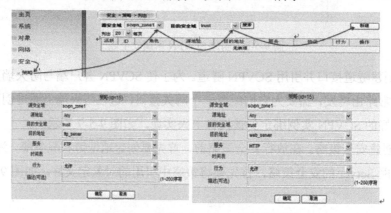

图 9-5-11 定义 Web Server 的地址对象

步骤 11：添加 SCVPN 用户账号，如图 9-5-12 所示。

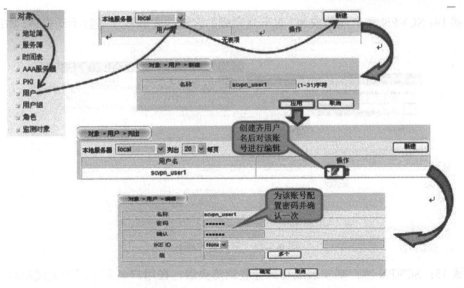

图 9-5-12　添加 SCVPN 用户账号

步骤 12：SCVPN 登录演示。在客户端上打开浏览器，在地址栏中输入"https://22 2.1.1.2:4433"，在登录界面中输入用户账号和密码，单击"登录"按钮，如图 9-5-13 所示。

图 9-5-13　登录界面

步骤 13：在初次登录时，会要求安装 SCVPN 客户端插件，此插件以 ActiveX 方式推送下载，并有可能被浏览器拦截，这时需要手动允许安装此插件，如图 9-5-14 所示。

图 9-5-14　安装插件

步骤14：SCVPN 客户端的安装，对下载完成的客户端安装程序进行手动安装，如图 9-5-15 所示。

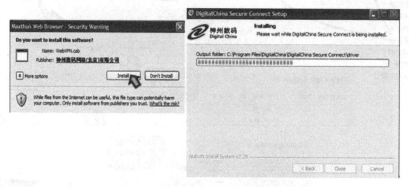

图 9-5-15　安装客户端软件

步骤15：SCVPN 客户端安装成功后会登录防火墙，在用户名和密码验证成功后，任务栏右下角的客户端程序图标会变成绿色，并在 Web 界面中显示"连接成功"，如图 9-5-16 所示。

图 9-5-16　连接成功界面

步骤16：单击任务栏中客户端程序图标，在弹出的菜单中选择"Network Information"选项，即可查看连接信息，如图 9-5-17 所示。

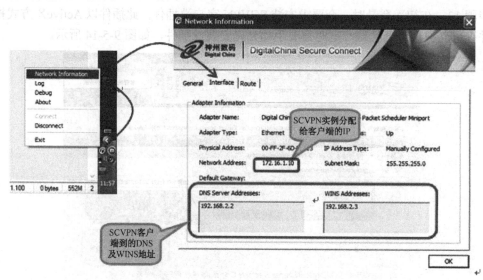

图 9-5-17　查看连接信息

步骤 17：在客户端的"Route"信息中查看下发给客户端的路由信息，如图 9-5-18 所示。

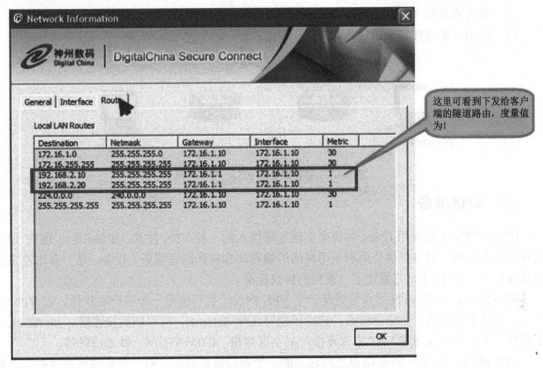

图 9-5-18　查看路由信息

♂ 小贴士

可以设置几个不同权限的角色，不同的账号所属的角色不同，这样不同的账号登录认证成功后，放行的服务不同或具有的权限不同。

任务六　建立路由器 IPSec VPN 隧道

♂ 需求分析

某公司由于规模的扩大，采用以前的 VPN 方法已经不能满足需求，公司内部网络通过一台路由器连接到公司外的另一台路由器上，出于安全性的考虑，网络管理员想更换一种更安全的 VPN 连接方式，网络管理员认为 IPSec VPN 可以实现此功能。

♂ 方案设计

IKE 技术提供了额外的特性，使配置 IPSec 更加灵活、更加容易。本任务通过在 Router1 和 Router2 上配置 VPN，实现网络之间通过 IPSec 的 VPN 方式保护总部与分部之间的数据通信。

所需设备如图 9-6-1 所示。

（1）DCR-2626 路由器 2 台。

（2）PC 2 台。
（3）交叉双绞线 3 条。
（4）路由器配置线缆 1 或 2 条。

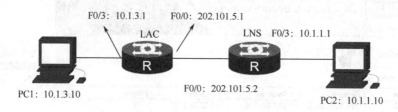

图 9-6-1　建立路由器 IPSec VPN 隧道

知识准备

IPSec VPN 是指采用 IPSec 协议来实现远程接入的一种 VPN 技术，IPSec 是由 IETF 定义的安全标准框架，用来提供公用和专用网络的端对端加密和验证服务。IPSec 是一套比较完整的成体系的 VPN 技术，它规定了一系列的协议标准。

VPN 结点：一个 VPN 结点可能是一个 VPN 网关，也可能是一个客户端软件。在 VPN 组网中，它属于组网的一个通信结点。它应该能够连接 Internet，有可能直接连接，如 ADSL、电话拨号等，也可能通过 NAT 方式连接，如小区宽带、CDMA 上网、铁通线路等。

VPN 隧道：在两个 VPN 结点之间建立的一个虚拟链路通道。两个设备内部的网络，能够通过这个虚拟的数据链路到达对方。与此相关的信息是当时两个 VPN 结点的 IP 地址、隧道名称、双方的密钥。

隧道路由：一个设备可能和很多设备建立隧道，那么就存在一个隧道选择的问题，即到什么目的地，走哪一条隧道。

IPSec 要解决的问题其实可以分为以下几个步骤。

（1）找到对方的 VPN 结点设备，如果对方是动态 IP 地址，那么必须能够通过一种有效途径及时发现对方 IP 地址的变化。

（2）建立隧道。如果两个设备都有合法的公网 IP 地址，那么建立一个隧道是比较容易的。一般通过内部的 VPN 结点发起一个 UDP 连接，再封装一次 IPSec，送到对方，因为 UDP 可以通过防火墙进行记忆，因此通过 UDP 封装的 IPSec 包，可以通过防火墙来回传递。

（3）建立隧道以后，需要确定隧道路由，即到哪里去，走哪条隧道。很多 VPN 隧道配置的时候就定义了保护网络，这样，隧道路由会根据保护的网络关系来决定。但是这样丧失了一定的灵活性。

常见的 IPSec VPN 类型有站到站、easy VPN（远程访问 VPN）、DMVPN（动态多点 VPN）、GET VPN（Group Encrypted Transport VPN）等。

任务实现

步骤 1：配置路由器 R1 的接口地址。

R1_config#interface fastEthernet 0/0
R1_config_f0/0#ip add 202.101.5.1 255.255.255.0
R1_config_f0/0#exit

```
R1_config#interface fastEthernet 0/3
R1_config_f0/3#ip add 10.1.3.1 255.255.255.0
R1_config_f0/3#exit
```

步骤 2：配置路由器 R2 的接口地址。

```
R2_config#interface fastEthernet 0/0
R2_config_f0/0#ip add 202.101.5.2 255.255.255.0
R2_config_f0/0#exit
R2_config#interface fastEthernet 0/3
R2_config_f0/3#ip add 10.1.1.1 255.255.255.0
R2_config_f0/3#exit
```

步骤 3：配置路由器 R1 的访问控制列表，定义需要 IPSec 保护的数据。

```
R1_config#ip access-list extended for_ipsec
R1_config_ext_nacl#permit ip 10.1.3.0 255.255.255.0 10.1.1.0 255.255.255.0
R1_config_ext_nacl#exit
```

步骤 4：配置路由器 R2 的访问控制列表，定义需要 IPSec 保护的数据。

```
R2_config#ip access-list extended for_ipsec
R2_config_ext_nacl#permit ip 10.1.1.0 255.255.255.0 10.1.3.0 255.255.255.0
R2_config_ext_nacl#exit
```

步骤 5：配置路由器 R1 的数据完整性验证和加密使用的变换集。

```
R1_config#crypto ipsec transform-set r1-r2
R1_config_crypto_trans#transform-type ah-sha-hmac esp-3des
R1_config_crypto_trans#mode tunnel
R1_config_crypto_trans#exit
```

步骤 6：配置路由器 R2 的数据完整性验证和加密使用的变换集。

```
R2_config#crypto ipsec transform-set r2-r1
R2_config_crypto_trans#transform-type ah-sha-hmac esp-3des
R2_config_crypto_trans#mode tunnel
R2_config_crypto_trans#exit
```

步骤 7：配置路由器 R1 加密映射表。

```
R1_config#crypto map r1 10 ipsec-isakmp
R1_config_crypto_map#match address for_ipsec
R1_config_crypto_map#set transform-set r1-r2
R1_config_crypto_map#set peer 202.101.5.2
R1_config_crypto_map#exit
```

步骤 8：配置路由器 R2 加密映射表。

```
R2_config#crypto map r2 10 ipsec-isakmp
R2_config_crypto_map#match address for_ipsec
R2_config_crypto_map#set transform-set r2-r1
R2_config_crypto_map#set peer 202.101.5.1
R2_config_crypto_map#exit
```

步骤 9：配置路由器 R1 生成最终密钥的预共享密钥。

```
R1_config#crypto isakmp key 12345 202.101.5.2
```

步骤 10：配置路由器 R2 生成最终密钥的预共享密钥。

R2_config#crypto isakmp key 12345 202.101.5.1

步骤 11：在 R1 接口中应用加密映射表。

R1_config#int f 0/0
R1_config_f0/0#crypto map r1

步骤 12：在 R2 接口中应用加密映射表。

R2_config#interface fastEthernet 0/0
R2_config_f0/0#crypto map r2

步骤 13：测试。注意，IPSec 的隧道建立需要触发条件，没有数据包流量时，是不会建立 SA（安全联盟）的。

```
R1#ping 10.1.1.1 -i 10.1.3.1         //注意,此处一定需要用源地址 ping 才能触发 IPSec 的形成
PING 10.1.1.1 (10.1.1.1): 56 data bytes
!!!!!
--- 10.1.1.1 ping statistics ---
5 packets transmitted, 5 packets received, 0% packet loss
round-trip min/avg/max = 0/2/10 ms
R1#sh crypto ipsec sa                //查看 IPSec 的安全联盟
Interface: FastEthernet0/0
Crypto map name:r1,   local addr. 202.101.5.1
  local    ident (addr/mask/prot/port): (10.1.3.0/255.255.255.0/0/0)
  remote ident (addr/mask/prot/port): (10.1.1.0/255.255.255.0/0/0)
  local crypto endpt.: 202.101.5.1,   remote crypto endpt.: 202.101.5.2
  inbound espsas:
    spi:0xf378f864(4084791396)
      transform:    esp-3des
      in use settings ={ Tunnel }
  sa timing: remaining key lifetime (k/sec):    (4607999/3592)
  inbound ah sas:
    spi:0xb27844(11696196)
      transform:    ah-sha-hmac
      in use settings ={ Tunnel }
  sa timing: remaining key lifetime (k/sec):    (4607999/3592)
  outbound espsas:
    spi:0x124596d7(306550487)
      transform:    esp-3des
      in use settings ={ Tunnel }
  sa timing: remaining key lifetime (k/sec):    (4607999/3592)
  outbound ah sas:
    spi:0xc756520f(3344323087)
      transform:    ah-sha-hmac
      in use settings ={ Tunnel }
  sa timing: remaining key lifetime (k/sec):    (4607999/3592)
R1#sh crypto isakmp sa                //查看 IKE SA
dstsrc           state          state-id         conn
202.101.5.2    202.101.5.1    <I>Q_SA_SETUP         2          4 r1 10
```

202.101.5.2	202.101.5.1	<I>M_SA_SETUP	1		4 r1 10

R1#
//查看 R2 对应的 IPSec 安全联盟和 IKE SA（第一阶段 SA）
R2#sh crypto ipsecsa
Interface: FastEthernet0/0
Crypto map name:r2 , local addr. 202.101.5.2
  local   ident (addr/mask/prot/port): (10.1.1.0/255.255.255.0/0/0)
  remote ident (addr/mask/prot/port): (10.1.3.0/255.255.255.0/0/0)
  local crypto endpt.: 202.101.5.2, remote crypto endpt.: 202.101.5.1
  inbound espsas:
    spi:0x124596d7(306550487)
      transform: esp-3des
      in use settings ={ Tunnel }
sa timing: remaining key lifetime (k/sec):   (4607999/3464)
  inbound ah sas:
    spi:0xc756520f(3344323087)
      transform: ah-sha-hmac
      in use settings ={ Tunnel }
sa timing: remaining key lifetime (k/sec):   (4607998/3464)
  outbound espsas:
    spi:0xf378f864(4084791396)
      transform: esp-3des
      in use settings ={ Tunnel }
sa timing: remaining key lifetime (k/sec):   (4607998/3464)
  outbound ah sas:
    spi:0xb27844(11696196)
      transform: ah-sha-hmac
      in use settings ={ Tunnel }
sa timing: remaining key lifetime (k/sec):   (4607998/3464)
R2#sh crypto isakmp sa

dstsrc		state	state-id	conn	
202.101.5.1	202.101.5.2	<R>Q_SA_SETUP	2		4 r2 10
202.101.5.1	202.101.5.2	<R>M_SA_SETUP	1		4 r2 10

R2#

任务七　防火墙 IPSec VPN 隧道的建立

♂ 需求分析

由于规模的扩大，总公司和分公司之间都购买了防火墙，两个公司之间出于安全性的考虑，要在两个防火墙之间使用一种更安全的 VPN 连接方式,网络管理员认为防火墙的 IPSec VPN 可以实现此功能。

♂ 方案设计

IPSec VPN 是现在互联网上最重要的网关到网关 VPN 技术,它已经成为企业分支机构间互连的首选。总部和分部要实现互访时,就会涉及此类 VPN,或者对数据包进行加密时会涉及 IPSec VPN。

所需设备如图 9-7-1 所示。
(1) 防火墙设备 2 台。
(2) 局域网交换机 n 台。
(3) 网线 n 条。
(4) PC n 台。

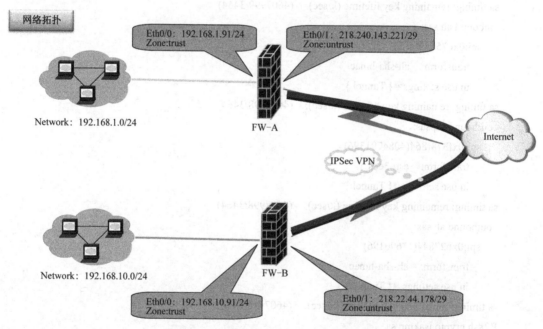

图 9-7-1 建立防火墙 IPSec VPN 隧道

任务要求:防火墙 FW-A 和 FW-B 都具有合法的静态 IP 地址,防火墙 FW-A 的内部保护子网为 192.168.1.0/24,防火墙 FW-B 的内部保护子网为 192.168.10.0/24。要求在 FW-A 与 FW-B 之间创建 IPSec VPN,使两端的保护子网能通过 VPN 隧道进行访问。

♂ 任务实现

步骤 1:创建 IKE 第一阶段提议,在 VPN/IKE VPN/P1 提议中定义 IKE 第一阶段的协商内容,两台防火墙的 IKE 第一阶段协商内容需要一致,如图 9-7-2 所示。

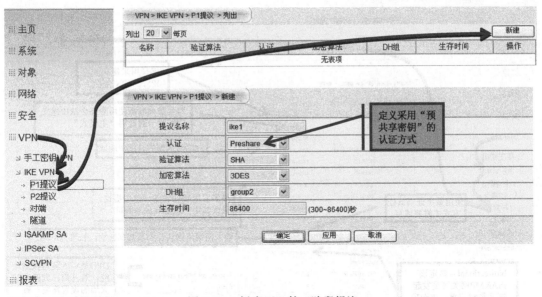

图 9-7-2　创建 IKE 第一阶段提议

步骤 2：创建 IKE 第二阶段提议，在 VPN/IKE VPN/P2 提议中定义 IKE 第二阶段的协商内容，两台防火墙的第二阶段协商内容需要一致，如图 9-7-3 所示。

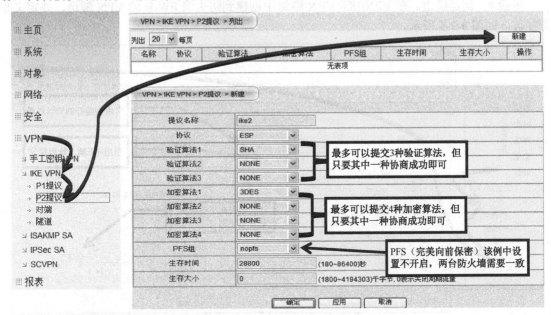

图 9-7-3　创建 IKE 第二阶段提议

步骤 3：创建对等体。在 VPN/IKE VPN/对端中创建"对等体"对象，并定义对等体的相关参数，如图 9-7-4 所示。

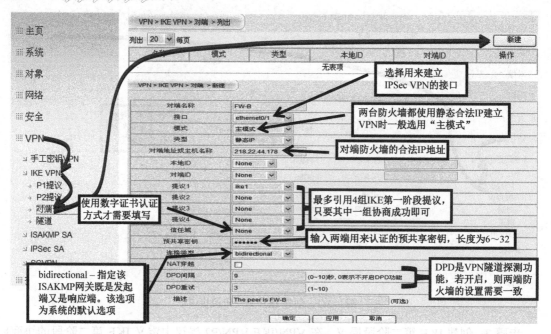

图 9-7-4 创建对等体

步骤 4：创建隧道。在 VPN/IKE VPN/隧道中创建到防火墙 FW-B 的 VPN 隧道，并定义相关参数，如图 9-7-5 所示。

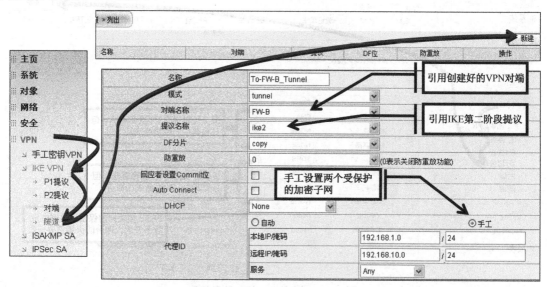

图 9-7-5 创建隧道

步骤 5：创建隧道接口并与 IPSec 绑定。在网络/接口中新建隧道接口，指定安全域并引用 IPSec 隧道，如图 9-7-6 所示。

步骤 6：添加隧道路由。在网络/路由/目的路由中新建一条路由，目的地址是对端加密保护子网，网关为创建的 tunnel 口，如图 9-7-7 所示。

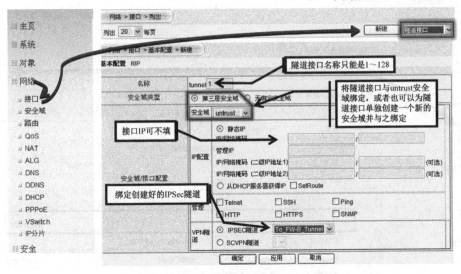

图 9-7-6 创建隧道接口并与 IPSec 绑定

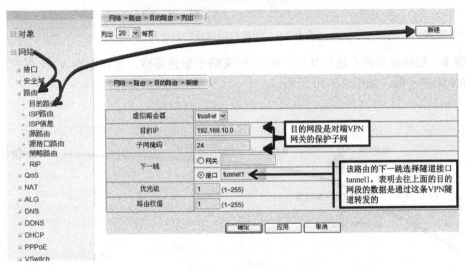

图 9-7-7 添加路由

步骤 7：添加安全策略。在创建安全策略前先要创建本地网段和对端网段的地址簿，如图 9-7-8 和图 9-7-9 所示。

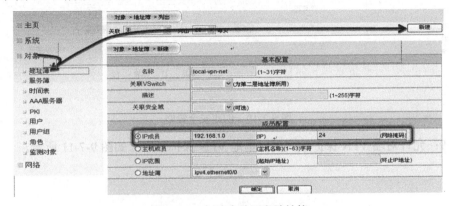

图 9-7-8 创建本地网段地址簿

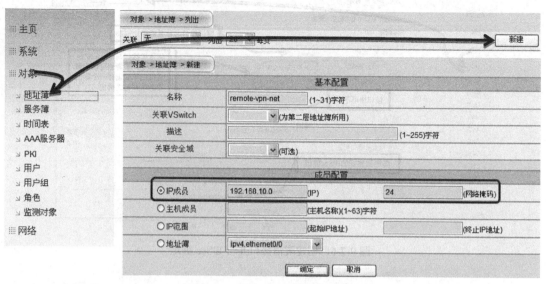

图 9-7-9　创建对端网段的地址簿

步骤 8：创建完成两个地址簿后，在安全/策略中新建策略，允许本地 VPN 保护子网访问对端 VPN 保护子网，如图 9-7-10 所示。

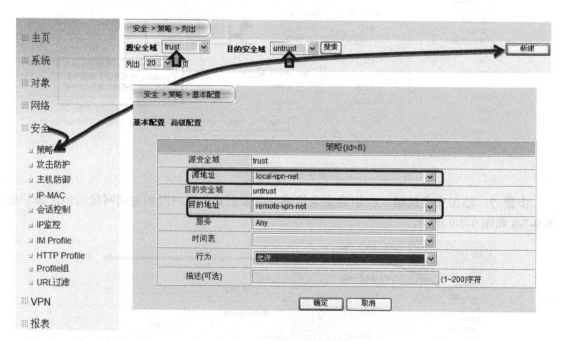

图 9-7-10　创建安全策略

步骤 9：允许对端 VPN 保护子网访问本地 VPN 保护子网，如图 9-7-11 所示。

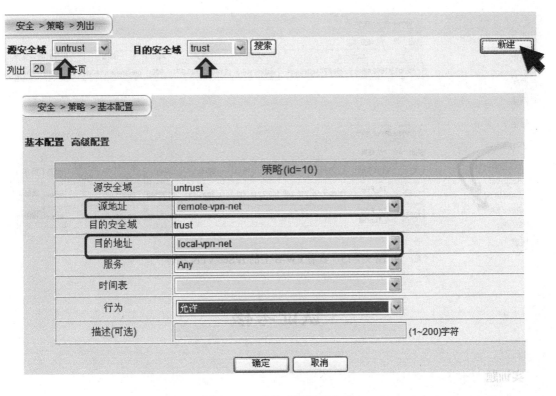

图 9-7-11 允许对端访问本地

步骤 10：FW-B 防火墙的配置步骤与 FW-A 相同，不同之处是某些步骤中的参数设置，此处略。

步骤 11：验证测试，查看防火墙 FW-A 上的 IPSec VPN 状态，如图 9-7-12 所示。

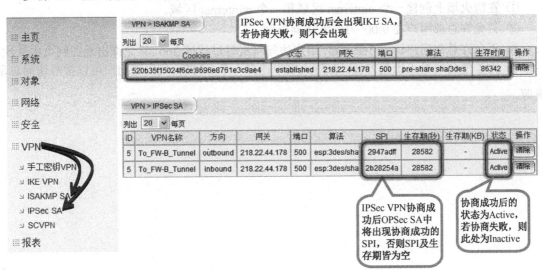

图 9-7-12 查看状态

步骤 12：查看防火墙 FW-B 上的 IPSec VPN 状态，如图 9-7-13 所示。

图 9-7-13　查看 FW-B 上的 IPSec VPN 状态

认证考核

实训题

（1）假设你是公司的网络管理员，公司因业务扩大，建立了一个分公司，公司的业务数据很重要，公司总部与分部传输数据时需要加密，采用 IPSec VPN 技术对数据进行加密。总部要加密的子网为 10.1.1.0/24，分部要加密的子网为 10.1.2.0/24。

（2）自己动手搭建网络环境，实现如下要求。

① 在防火墙上创建一个 ninternet 区域和一个 winternet 区域。

② 只允许总部用户在 9:00～17:30 访问外网。

③ 不允许总部用户访问 360.com，用户访问网页时过滤掉 360，只允许出现在网页中 3 次。

④ 为了使公司员工出差方便，在防火墙中建立一个 L2TP，使出差用户可以拨号访问总部资源。

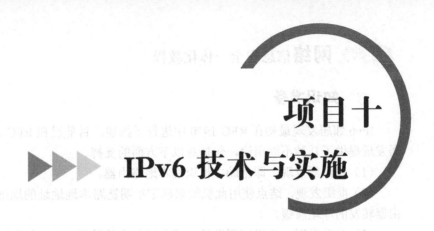

项目十

IPv6 技术与实施

教学背景

IPv6 协议是取代 IPv4 的下一代网络协议,它具有许多新的特性与功能。由 IP 地址危机产生和发展起来的 IPv6 作为下一代互联网协议已经得到了各方的公认,未来互联网的发展离不开 IPv6 的支持和应用。

任务一 IPv6 邻居发现

需求分析

某公司使用了 IPv6 网络,网络管理员想使路由器能够相互学习,决定启用 IPv6 的邻居发现,这样路由器就可以轻松地进行通信了。

方案设计

在两台 IPv6 路由设备上启用 IPv6 routing,用双绞线互连两台设备,使能用 IPv6,然后查看邻居表,互相使用命令 ping6 测试与对方的 IPv6 接口地址的连通性,也可采用 tracert6 命令进行验证。

所需设备如图 10-1-1 所示。

（1）路由器 2 台。
（2）Console 线 1 条。
（3）PC 1 台。
（4）网线若干。

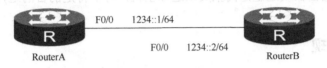

图 10-1-1 IPv6 邻居发现

任务要求：按照拓扑图连接网络,RouterA 的 Console 口接上 Console 线,并与 PC 相连；在配置 RouterB 的时候,将 Console 线接到 RouterB 的 Console 口上。在端口上配置 IPv6 地址,通过 IPv6 邻居发现命令查看各项功能的实现过程,通过 IPv6 邻居发现命令查看邻居缓存。

知识准备

IPv6 邻居发现最初在 RFC 1970 中进行了描述,目前已在 RFC 2461 中重新定义。IPv6 邻居发现提供了几种不同用途,包括对以下方面的支持。

(1) 路由器发现,即帮助主机识别本地路由器。

(2) 前缀发现,结点使用此机制来确定指明链路本地地址的地址前缀,以及必须发送给路由器转发的地址前缀。

(3) 参数发现,此机制帮助结点确定诸如本地链路 MTU 之类的信息。

(4) 地址自动配置,用于 IPv6 结点地址自动配置。

(5) 地址解析,替代了 ARP 和 RARP,帮助结点从目的 IP 地址中确定本地结点(即邻居)的链路层地址。

(6) 下一跳确定,可用于确定包的下一个目的地,即可确定包的目的地是否在本地链路上。如果在本地链路上,则下一跳就是目的地;否则,包需要路由,下一跳就是路由器,邻居发现可用于确定应使用的路由器。

(7) 邻居不可达检测,邻居发现可帮助结点确定邻居(目的结点或路由器)是否可达。

(8) 重复地址检测,邻居发现可用于帮助结点确定它想使用的地址在本地链路上是否已被占用。

(9) 重定向。有时结点选择的转发路由器对于待转发的包而言并非最佳,这种情况下,该转发路由器可以对结点进行重定向,以将包转发到最佳的路由器中。例如,结点将发往 Internet 的包发送给为结点的内联网服务的默认路由器,该内联网路由器可以对结点进行重定向,以将包发送给连接在同一本地链路上的 Internet 路由器。

邻居发现服务通过 5 种 ICMPv6 报文类型来执行,这些报文如下所示。

(1) 路由器宣告。要求路由器周期性地发送多点传送路由器宣告消息,宣告其可用性及其可到达的在线结点、用于配置的链路和 Internet 参数。这些宣告包含对所使用的网络地址前缀、建议的路程段极限值及本地的 MTU 的指示,也包括指明结点应使用的自动配置类型的标志。

(2) 路由器请求。主机可以请求本地路由器立即发送其路由器宣告。路由器必须周期性地发送这些宣告,但是在收到路由器请求报文时,不必等到下一个预定传送时间到达,而应立即发送宣告信息。

(3) 邻居宣告。结点在收到邻居请求报文时或者链路层地址改变时,发出邻居宣告报文。

(4) 邻居请求。结点发送邻居请求来请求邻居的链路层地址,以验证它先前所获得并保存在缓存中的邻居链路层地址的可达性,或者验证它自己的地址在本地链路上是唯一的。

(5) 重定向。路由器发送重定向报文以通知主机,对于特定的目的地,自己不是最佳的路由器。

任务实现

步骤 1:在路由器端口上配置 IPv6 地址。

```
RouterA#debug ipv6 nd           //打开 IPv6 邻居发现的 debug 开关
ICMP6-ND debugging is on
RouterA#config
RouterA_config#int f0/0         //在端口上配置 IPv6 地址
```

项目十 IPv6 技术与实施

RouterA_config_f0/0#ipv6 address 1234::1/64
//输出的调试信息：IPv6 邻居发现进行重复地址检测的过程
RouterA_config_f0/0#2002-1-1 00:23:33 ICMP6-ND: sending NS for FF02::1:FF26:C050 on FastEthernet0/0
2002-1-1 00:23:33 ICMP6-ND: sending NS for FF02::1:FF00:1 on FastEthernet0/0
2002-1-1 00:23:34 FastEthernet0/0: DAD complete for 1234::1 no duplicates found
2002-1-1 00:23:34 ICMP6-ND: sending NA for 1234::1 on FastEthernet0/0.
2002-1-1 00:23:34 ICMP6-ND: adding prefix 1234::1/64 to FastEthernet0/0.
2002-1-1 00:23:34 FastEthernet0/0: DAD complete for FE80:4::2E0:FFF:FE26:C050 - no duplicates found
2002-1-1 00:23:35 ICMP6-ND: sending NA for FE80::2E0:FFF:FE26:C050 on FastEthernet0/0.
RouterB_config_f0/0#ipv6 address 1234::2/64   //同样，在 RouterB 上配置 IPv6 地址

步骤 2：在路由器上验证配置。

RouterB#show ipv6 int f0/0
FastEthernet0/0 is up, line protocol is up
  IPv6 is enabled, link-local address is FE80::2E0:FFF:FE27:4230
  Global unicast address(es):
    1234::2, subnet is 1234::/64
  Joined group address(es):
    FF02::1
    FF02::1:FF00:2
    FF02::1:FF27:4230
  MTU is 1500 bytes
  ICMP error messages limited to one every 100 milliseconds
  ICMP redirects are enabled
  ICMP unreachables are enabled
RouterB#

步骤 3：在 RouterA 上使用命令 ping6 验证与 RouterB 的连通性。

RouterA#ping6 1234::2
PING 1234::2 (1234::2): 56 data bytes
!!!!!
--- 1234::2 ping6 statistics ---
5 packets transmitted, 5 packets received, 0% packet loss
round-trip min/avg/max = 0/0/0 ms
//下面为 IPv6 邻居发现的调试信息，显示了进行地址解析及邻居可达性检测的过程
RouterA#2002-1-1 00:32:50 DELET -> INCOM: 1234::2
//发送 NS，检测是否有此邻居（1234::2），同时解析此邻居的链路层地址
2002-1-1 00:32:50 ICMP6-ND: sending NS for FF02::1:FF00:2 on FastEthernet0/0
/*收到 NA，得知该邻居存在并可达，邻居可达性检测成功；从此 NA 消息中获得了该邻居的链路层地址，地址解析成功*/
2002-1-1 00:32:50 ICMP6-ND: Received NA for 1234::2 on FastEthernet0/0 from 1234::2.
2002-1-1 00:32:50 INCOM -> REACH: 1234::2
RouterA#

验证可以通过查看邻居缓存来实现。

步骤 4：查看邻居缓存。

RouterA#show ipv6 neighbors

IPv6 Address	Age Link-layer AddrState	Interface
1234::2	43 00:e0:0f:27:42:30 REACH	FastEthernet0/0

步骤 5：配置路由器的路由转发功能。

```
RouterA_config#ipv6 unicast-routing            //配置 RouterA
RouterB_config#ipv6 unicast-routing            //配置 RouterB
```

步骤 6：关闭 debug 开关。

```
RouterA#nodebug ipv6 nd
```

步骤 7：验证结果。

验证路由器发现和前缀发现的功能，可以从 PC 连接到路由器上来进行实验。

任务二　IPv6 ISATAP 隧道搭建

♂ 需求分析

某公司准备对公司网络进行升级，引入新的网络架构——采用 IPv6 来建立公司的网络，总公司和分公司都已经成功建立了基于 IPv6 的内部网络，但是总公司和分公司需要联系的时候仍然需要通过 IPv4 网络。

♂ 方案设计

在两台路由器上分别配置 ISATAP 隧道，两台路由器之间通过 IPv4 网络连接。
所需设备如图 10-2-1 所示。
（1）路由器 2 台。
（2）Console 线 1 条。
（3）PC 1 台。
（4）网线若干。

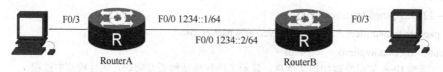

图 10-2-1　IPv6 ISATAP 隧道搭建

♂ 知识准备

ISATAP（Intra-Site Automatic Tunnel Addressing Protocol，站内自动隧道寻址协议）是一种非常容易部署和使用的 IPv6 过渡机制。在一个 IPv4 网络中，可以非常轻松地进行 ISATAP 的部署，首先 PC 必须是 v4/v6 双栈 PC，并且需要有一台支持 ISATAP 的路由器，ISATAP 路由器可以在网络中的任何位置，只要 PC 能够 ping 通即可。可以在路由器上部署 ISATAP，这种网络中支持 ISATAP 的双栈主机，在需要访问 IPv6 资源时，可以与 ISATAP 路由器建立 ISATAP 隧道，ISATAP 主机根据 ISATAP 路由器发送的 IPv6 前缀构造自己的 IPv6 地址（此 IPv6 地址被自动关联到 ISATAP 主机本地产生的一个 ISATAP 虚拟网卡上），并且将这台 ISATAP 路由器设置为自己的 IPv6 默认网关，这台主机就能够通过 ISATAP 路由器访问 IPv6 的资源了。

这种方法部署起来非常简单，在许多场合下，客户为了节省成本，又希望网络中的 IPv6 主机能够访问其资源，又不愿意对现有网络做大规模的变更及设备升级，就可以采用这种方法，即购买一台支持 ISATAP 的路由器，甚至可以将 ISATAP 路由器旁挂在网络上，只要它能够访问 v6 资源并且响应 ISATAP PC 的隧道建立请求即可。

任务实现

步骤 1：在 RouterA 上配置 ISATAP 隧道。

```
RouterA_config#interface tunnel 0
RouterA_config_t0# ipv6 address 1234::/16 eui-64
RouterA_config_t0# tunnel mode ipv6 ipisatap
RouterA_config_t0# tunnel source 1.1.1.1
RouterA_config_t0# exit
RouterA_config# interface fastEthernet 0/0
RouterA_config_f0/0#ip address 1.1.1.1 255.255.255.0
RouterA_config_f0/0#exit
RouterA_config#
```

验证：

```
RouterA_config#show ipv6 int
Tunnel0 is up, line protocol is up
  IPv6 is enabled, link-local address is FE80::5EFE:101:101
  Global unicast address(es):
    1234::5EFE:101:101, subnet is 1234::/16 [EUI]
  Joined group address(es):
    FF02::1
    FF02::1:FF01:101
  MTU is 1460 bytes
  ICMP error messages limited to one every 100 milliseconds
  ICMP redirects are enabled
  ICMP unreachables are enabled
RouterA_config#
```

步骤 2：在 RouterB 上配置 ISATAP 隧道。

```
RouterB_config#interface tunnel 0
RouterB_config_t0# ipv6 address 1234::/16 eui-64
RouterB_config_t0# tunnel mode ipv6 ipisatap
RouterB_config_t0# tunnel source 1.1.1.2
RouterA_config_t0# exit
RouterA_config# interface fastEthernet 0/0
RouterB_config_f0/0#ip address 1.1.1.2 255.255.255.0
RouterB_config_f0/0#exit
RouterB_config#ipv6 route default Tunnel0
RouterB_config#
```

验证：

```
RouterB_config#show ipv6 int
Tunnel0 is up, line protocol is up
    IPv6 is enabled, link-local address is FE80::5EFE:101:102
    Global unicast address(es):
        1234::5EFE:101:102, subnet is 1234::/16 [EUI]
    Joined group address(es):
        FF02::1
        FF02::1:FF01:102
    MTU is 1460 bytes
    ICMP error messages limited to one every 100 milliseconds
    ICMP redirects are enabled
    ICMP unreachables are enabled
RouterB_config#
```

步骤 3：验证测试。

```
RouterA_config#ping6 1234::5EFE:101:102
PING 1234::5EFE:101:102 (1234::5EFE:101:102): 56 data bytes
!!!!!
--- 1234::5EFE:101:102 ping6 statistics ---
5 packets transmitted, 5 packets received, 0% packet loss
round-trip min/avg/max = 0/2/10ms
RouterA_config#
RouterB_config#ping6 1234::5EFE:101:101
PING 1234::5EFE:101:101 (1234::5EFE:101:101): 56 data bytes
!!!!!
--- 1234::5EFE:101:101 ping6 statistics ---
5 packets transmitted, 5 packets received, 0% packet loss
round-trip min/avg/max = 0/0/0 ms
RouterB_config#
```

♂ 小贴士

在任务开始前，路由器为默认配置；两个路由器中如果有 IPv4 网络，则应保证 IPv4 网络畅通。

任务三　实现 6 to 4 隧道

♂ 需求分析

某公司准备对公司网络进行升级，引入新的网络架构——采用 IPv6 来建立公司的网络，总公司和分公司都已经成功建立了基于 IPv6 的内部网络，但是总公司和分公司需要联系的时候仍然需要通过 IPv4 网络。

项目十 IPv6 技术与实施

♂ 方案设计

在两台路由器上分别配置 IPv6 to IPv4 手工隧道，两台路由器之间通过 IPv4 网络连接。
所需设备如图 10-3-1 所示。
（1）路由器 2 台。
（2）Console 线 1 条。
（3）PC 1 台。
（4）网线若干。

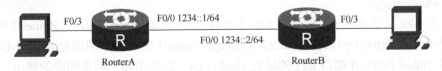

图 10-3-1　实现 6 to 4 隧道

♂ 知识准备

6 to 4 是一种自动构造隧道的方式，它的好处在于只需要一个全球唯一的 IPv4 地址便可使整个站点获得 IPv6 的连接。国际地址分配机构专门为 6 to 4 隧道分配了一个永久性的 13 比特的顶级聚类标识（TLA ID），相应的网络前缀是 2002::/16。利用 6 to 4 地址，隧道末端的 IPv4 地址可以从目的 IPv6 地址的 48 比特前缀中自动提取出来，地址前缀后面的部分 SLA ID+接口 ID 唯一地标识了该主机在站点中的位置，假设 IPv4 地址为 138.14.85.210，转换成十六进制 8a0e:55d2，则 6 to 4 地址前缀为 2002:8a0e:55d2::/48。

6 to 4 采用特殊的 IPv6 地址使 IPv6 孤岛相互连接。此时，IPv6 的出口路由器与其他的 IPv6 域建立了隧道连接。IPv4 隧道的末端可从 IPv6 域的地址前缀中自动提取，通过这个机制，站点能够配置 IPv6 而不需要向注册机构申请 IPv6 地址空间。这也简化了 ISP 的管理工作。可以设想，在一个拥有很多部门的企业中，各部门内部使用私有地址和 NAT 技术，利用 6 to 4 策略可以建立一个虚拟 IPv6 外部网。它同样可以重新建立点到点的 IP 连接，且允许企业在不同地方的服务器上使用 IPSec 协议，从而进一步提高了网络的安全性。此外，6 to 4 机制还允许在采用 6 to 4 的 IPv6 站点和纯 IPv6 站点之间通过中继路由器进行通信，此时不要求通信的两个端点之间具有可用的 IPv4 连接，中继路由器建议运行 BGP4+，使其应用范围更广。

♂ 任务实现

步骤 1：在 RouterA 上配置 IPv6 to IPv4 隧道。

```
RouterA_config#interface tunnel 0
RouterA_config_t0# ipv6 address 2002:101:101::1/32
RouterA_config_t0# tunnel mode ipv6ip 6to4
RouterA_config_t0# tunnel source 1.1.1.1
RouterA_config_t0# interface fastEthernet 0/0
RouterA_config_f0/0#ip address 1.1.1.1 255.255.255.0
RouterA_config_f0/0#exit
RouterA_config#
```

步骤 2：在 RouterB 上配置 IPv6 to IPv4 隧道。

```
RouterB_config#interface tunnel 0
RouterB_config_t0# ipv6 address 2002:101:102::1/32
RouterB_config_t0# tunnel mode ipv6ip 6to4
RouterB_config_t0# tunnel source 1.1.1.2
RouterB_config_t0# exit
RouterB_config# interface fastEthernet 0/0
RouterB_config_f0/0#ip address 1.1.1.2 255.255.255.0
RouterB_config_f0/0#exit
RouterB_config#
```

步骤 3：测试。在 RouterA 的端口 f0/1 上配置 IPv6 地址 abcd::1/64，在 RouterB 的端口 f0/1 上配置 IPv6 地址 dcba::1/64。在 RouterA 上 ping6 RouterB 端口 f0/1 的地址 dcba::1/64，或者在 RouterB 上 ping6 RouterA 端口 f0/1 的地址 abcd::1/64。在路由器上配置相应的路由。

```
RouterA_config# interface fastEthernet 0/1
RouterA_config_f0/1#
RouterA_config#ipv6 route dcba::/61 2002:101:102::1
RouterB_config# interface fastEthernet 0/1
RouterB_config_f0/1#
RouterB_config# ipv6 route abcd::/64 2002:101:101::1
```

步骤 4：在 RouterA 上验证配置。

```
RouterA# ping6 dcba::1
PING dcba::1 (DCBA::1): 56 data bytes
!!!!!
--- dcba::1 ping6 statistics ---
5 packets transmitted, 5 packets received, 0% packet loss
round-trip min/avg/max = 0/0/0 ms
RouterA#
```

步骤 5：在 RouterB 上验证配置。

```
RouterB#ping6 abcd::1
PING abcd::1 (ABCD::1): 56 data bytes
!!!!!
--- abcd::1 ping6 statistics ---
5 packets transmitted, 5 packets received, 0% packet loss
round-trip min/avg/max = 0/0/0 ms
RouterB#
```

小贴士

在任务开始前，路由器为默认配置；保证两个路由器的端口 f0/1 状态都为 up；两个路由器中间如果有 IPv4 网络，则应保证 IPv4 网络畅通。

任务四　IPv6 RIPng 配置

需求分析

网络管理员接到任务，准备对公司网络进行升级，引入新的网络架构——采用 IPv6 来建立公司的网络，网络管理员分析后，决定采用 IPv6 RIPng，通过配置完成了网络的连接。

方案设计

在 3 台路由器上启用 IPv6 routing，配置好 IPv6 地址后，即可对路由器进行 RIPng 配置。所需设备如图 10-4-1 所示。

（1）路由器 3 台。
（2）串口线 2 条，转接线 1 条。
（3）网线若干。
（4）Console 线 1 条。
（5）PC 1 台。

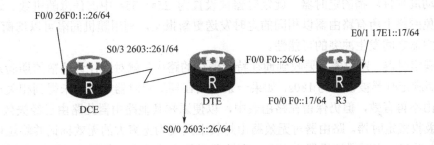

图 10-4-1　IPv6 RIPng 配置

知识准备

RIPng（RIP for IPv6）下一代路由选择信息协议是一种基于 IPv6 网络协议和算法的协议。在国际性网络中，如因特网，拥有很多应用于整个网络的路由选择协议。形成网络的每一个自治系统都有属于自己的路由选择技术，不同的自治系统，路由选择技术也不同。自治系统内部的路由选择协议称为内部网关协议（IGP）。外部网关协议（EGP）是一种用于在自治系统之间传输路由选择信息的协议。RIPng 在中等规模的自治系统中被用作 IGP。对于较复杂的网络环境，RIPng 不适用。

RIPng 是一种距离矢量算法，RIP 既适用于 XNS 协议，又适用于路由类协议。

1. RIPng 报文格式

RIPng 是基于 UDP 的协议，并且使用端口 521 发送和接收数据报。RIPng 报文大致可分为两类：选路信息报文和用于请求信息的报文。它们都使用相同的格式，由固定的首部和路由表项组成，其中，路由表项可以有多个。

2. RIPng 工作原理

路由器通常不会主动发出请求报文来进行路由请求，通常只在路由器刚启动或路由器正在寻找路由信息时才会发出请求报文以获得响应。路由器在查询响应、周期更新、触发更新3种情况下会收到响应报文。路由器根据响应报文判断是否对本地路由表进行更新。由于响应报文可能对本地路由表进行改动，因此必须对报文的来源进行严格的检查，以确认报文的合法性。

众所周知，基于距离矢量算法的路由协议会产生慢收敛和无限计数问题，这样就引发了路由的不一致。RIPng 使用水平分割技术、毒性逆转技术、触发更新技术来解决这些问题，但是这些技术的引入，又带来了另外一些问题，如采用触发更新技术后，如果不对产生的报文进行合理的控制，很容易产生广播风暴。

路由器周期性的报文广播和触发更新给网络造成了很多额外的负载，为减少路由信息的数量，RIPng 可以采用多播技术发送更新报文，并利用一个小的随机时延对触发更新报文进行抑制。

定时器在 RIPng 中起着非常重要的作用，RIPng 使用定时器来实现路由表的更新、报文的发送。周期性的报文广播是由定时器实现的。为防止路由表长时间未更新而失效，每个路由表项有两个定时器与之相联系，超时的路由表项最终会被删除，以防止路由器广播和使用已经失效的路由。RIPng 中使用的定时器主要有以下 3 个。

（1）启动周期性广播的定时器。此定时器被设置为 25s~35s 中的任意随机数。这样设置的目的是避免网络上所有路由器以相同的定时发送更新报文，利用随机间隔可以均衡通信量，从而减少路由器之间发生冲突的可能性。

（2）期满定时器。路由器只要收到通往特定信宿的路由，就对通往该信宿的期满定时器进行初始化。期满定时器被设定为 180s，如果一条路由在期满定时器超时前未得到相关报文的更新，则该路由不再有效，但仍保留在路由表中，以便通知其他路由器此路由已经失效。

（3）垃圾收集定时器。路由器对无效路由封装上尺度为无穷大的无效标记并将垃圾收集定时器初始化。此时，定时器被设置为 120s，在这段时间内这些路由仍然会被路由器周期性地广播，这样相邻路由器就能迅速从路由表中删除该路由了。

♂ 任务实现

步骤 1：DCE 路由器基本配置。

```
DCE_config#ipv6 unicast-routing
DCE_config_f0/0#ipv6 enable
DCE_config_f0/0#ipv6 address 26f0:1::26/64
DCE_config_s0/3#ipv6 enable
DCE_config_s0/3#encap ppp
DCE_config_s0/3#physical speed 64000
DCE_config_s0/3#ipv6 address 2603::261/64
```

步骤 2：验证配置。

```
Router#show ipv6 interface f0/0
FastEthernet0/0 is up, line protocol is up
    IPv6 is enabled, link-local address is FE80::2E0:FFF:FE7A:840
    Global unicast address(es):
```

         26F0:1::26, subnet is 26F0:1::/64
      Joined group address(es):
         FF02::1
         FF02::2
         FF02::1:FF00:26
         FF02::1:FF7A:840
      MTU is 1500 bytes
      ICMP error messages limited to one every 100 milliseconds
      ICMP redirects are enabled
      ICMP unreachables are enabled
Router#show ipv6 interface s0/3
Serial0/3 is up, line protocol is up
      IPv6 is enabled, link-local address is FE80::2E0:FFF:FE7A:840
      Global unicast address(es):
         2603::261, subnet is 2603::/64
      Joined group address(es):
         FF02::1
         FF02::2
         FF02::1:FF00:261
         FF02::1:FF7A:840
      MTU is 1500 bytes
      ICMP error messages limited to one every 100 milliseconds
      ICMP redirects are enabled
      ICMP unreachables are enabled
Router#sho ipv6 route
C   2603::/64[2]
is directly connected, C,Serial0/3
C   2603::261/128[2]
is directly connected, L,Serial0/3
C   26f0:1::/64[1]
is directly connected, C,FastEthernet0/0
C   26f0:1::26/128[1]
is directly connected, L,FastEthernet0/0
C   fe80::/10[2]
is directly connected, L,Null0
C   fe80::/64[2]
is directly connected, C,Serial0/3
C   fe80::2e0:fff:fe7a:840/128[2]
is directly connected, L,Serial0/3

步骤 3：DTE 路由器基本配置。

DTE_config#ipv6 unicast-routing
DTE_config_f0/0#ipv6 enable
DTE_config_f0/0#ipv6 address f0::26/64
DTE_config_s0/3#ipv6 enable

```
DTE_config_s0/3#encap ppp
DTE_config_s0/3#ipv6 address 2603::26/64
```

步骤4：验证配置。

```
Router#show ipv6 interface f0/0
FastEthernet0/0 is up, line protocol is up
    IPv6 is enabled, link-local address is FE80::2E0:FFF:FE7A:A00
    Global unicast address(es):
        F0::26, subnet is F0::/64
    Joined group address(es):
        FF02::1
        FF02::2
        FF02::1:FF00:26
        FF02::1:FF7A:A00
    MTU is 1500 bytes
    ICMP error messages limited to one every 100 milliseconds
    ICMP redirects are enabled
    ICMP unreachables are enabled
Router#show ipv6 interface s0/3
Serial0/3 is up, line protocol is up
    IPv6 is enabled, link-local address is FE80::2E0:FFF:FE7A:A00
    Global unicast address(es):
        2603::26, subnet is 2603::/64
    Joined group address(es):
        FF02::1
        FF02::2
        FF02::1:FF00:26
        FF02::1:FF7A:A00
    MTU is 1500 bytes
    ICMP error messages limited to one every 100 milliseconds
    ICMP redirects are enabled
    ICMP unreachables are enabled
Router#sho ipv6 route
C       f0::/64[1]
is directly connected, C,FastEthernet0/0
C       f0::26/128[1]
is directly connected, L,FastEthernet0/0
C       2603::/64[1]
is directly connected, C,Serial0/3
C       2603::26/128[1]
is directly connected, L,Serial0/3
C       fe80::/10[3]
is directly connected, L,Null0
C       fe80::/64[1]
is directly connected, C,Serial0/3
```

```
    C       fe80::2e0:fff:fe7a:a00/128[1]
is directly connected, L,Serial0/3
    C       fe80::/64[1]
is directly connected, C,FastEthernet0/0
    C       fe80::2e0:fff:fe7a:a00/128[1]
is directly connected, L,FastEthernet0/0
```

步骤 5：R3 路由器的配置。

```
R3_config#ipv6 unicast-routing
R3_config_f0/0#ipv6 enable
R3_config_f0/0#ipv6 address f0::17/64
R3_config_e0/1#ipv6 address 17e1::17/64
```

步骤 6：验证配置。

```
Router#show ipv6 interface f0/0
FastEthernet0/0 is up, line protocol is up
  IPv6 is enabled, link-local address is FE80::2E0:FFF:FE9B:7CF1
  Global unicast address(es):
    F0::17, subnet is F0::/64
  Joined group address(es):
    FF02::1
    FF02::2
    FF02::1:FF00:17
    FF02::1:FF9B:7CF1
  MTU is 1500 bytes
  ICMP error messages limited to one every 100 milliseconds
  ICMP redirects are enabled
  ICMP unreachables are enabled
Router#show ipv6 interface e0/1
Ethernet0/1 is up, line protocol is up
  IPv6 is enabled, link-local address is FE80::2E0:FFF:FE9B:7CF2
  Global unicast address(es):
    17E1::17, subnet is 17E1::/64
  Joined group address(es):
    FF02::1
    FF02::2
    FF02::1:FF00:17
    FF02::1:FF9B:7CF2
  MTU is 1500 bytes
  ICMP error messages limited to one every 100 milliseconds
  ICMP redirects are enabled
  ICMP unreachables are enabled
Router#show ipv6 route
    C       f0::/64[2]
is directly connected, C,FastEthernet0/0
    C       f0::17/128[2]
```

```
               is directly connected, L,FastEthernet0/0
    C          17e1::/64[2]
               is directly connected, C,Ethernet0/1
    C          17e1::17/128[2]
               is directly connected, L,Ethernet0/1
    C          fe80::/10[2]
               is directly connected, L,Null0
    C          fe80::/64[2]
               is directly connected, C,FastEthernet0/0
    C          fe80::2e0:fff:fe9b:7cf1/128[2]
               is directly connected, L,FastEthernet0/0
    C          fe80::/64[2]
               is directly connected, C,Ethernet0/1
    C          fe80::2e0:fff:fe9b:7cf2/128[2]
               is directly connected, L,Ethernet0/1
```

步骤 7：DCE 路由器的 RIPng 基本配置。

生成 RIPng 实例，名称为 r1（可以任取），成功后进入 r1 的配置状态。

```
Router_config#ipv6 router rip r1
```

转发静态路由，产生 RIPng 路由。此处可以进行其他的 RIPng 基本配置。

```
Router_config_ripng_r1#redis connect
```

在 S0/3 口上 enable 实例 r1：

```
Router_config_s0/3#ipv6 rip r1 enable
```

步骤 8：DTE 路由器的 RIPng 基本配置。

生成 RIPng 实例，名称为 r1（可以任取）。

```
Router_config#ipv6 router rip r1
```

转发静态路由，产生 RIPng 路由。

```
Router_config_ripng_r1#redis connect
```

在 F0/0、S0/3 口上 enable 实例 r1：

```
Router_config_f0/0#ipv6 rip r1 enable
Router_config_s0/3#ipv6 rip r1 enable
```

步骤 9：R3 路由器的 RIPng 基本配置。

生成 RIPng 实例，名称为 r1（可以任取）。

```
Router_config#ipv6 router rip r1
```

转发静态路由，产生 RIPng 路由。

```
Router_config_ripng_r1#redis connect
```

在 F0/0 口上 enable 实例 r1：

```
Router_config_f0/0#ipv6 rip r1 enable
```

步骤 10：DCE 基本配置验证。

```
Router#sho ipv6 rip r1 da
  f0::/64
                  [2,120]    via fe80::2e0:fff:fe7a:a00 (on Serial0/3)      N, from:
[fe80::2e0:fff:fe7a:a00]
  17e1::/64
```

```
            [3,120]    via fe80::2e0:fff:fe7a:a00 (on Serial0/3)      N, from
[fe80::2e0:fff:fe7a:a00]
 2603::/64
            [1,120]    via if 3      R   from:[1,0]
 26f0:1::/64
            [1,120]    via if 4      R   from:[1,0]
Router#show ipv6 route
Codes: C - Connected, L - Local, S - Static, R - Ripng, B - BGP
       ON1 - OSPF NSSA external type 1, ON2 - OSPF NSSA external type 2
       OE1 - OSPF external type 1, OE2 - OSPF external type 2
       DHCP - DHCP type
VRF ID: 0
R      f0::/64[1]
       [120,2] via fe80::2e0:fff:fe7a:a00(onSerial0/3)
R      17e1::/64[1]
       [120,3] via fe80::2e0:fff:fe7a:a00(onSerial0/3)
C      2603::/64[2]
is directly connected, C,Serial0/3
C      2603::261/128[2]
is directly connected, L,Serial0/3
C      26f0:1::/64[2]
is directly connected, C,FastEthernet0/0
C      26f0:1::26/128[2]
is directly connected, L,FastEthernet0/0
…
```

步骤 11：DTE 基本配置验证。

```
Router_config#sho ipv6 rip r1 da
 f0::/64
            [1,120]    via if 4      R   from:[1,0]
 17e1::/64
            [2,120]    via fe80::2e0:fff:fe9b:7cf1 (on FastEthernet0/0)   N, from:
[fe80::2e0:fff:fe9b:7cf1]
 2603::/64
            [1,120]    via if 3      R   from:[1,0]
 26f0:1::/64
            [2,120]    via fe80::2e0:fff:fe7a:840 (on Serial0/3)       N, from:
[fe80::2e0:fff:fe7a:840]

Router#show ipv6 route
Codes: C - Connected, L - Local, S - Static, R - Ripng, B - BGP
       ON1 - OSPF NSSA external type 1, ON2 - OSPF NSSA external type 2
       OE1 - OSPF external type 1, OE2 - OSPF external type 2
       DHCP - DHCP type
VRF ID: 0
```

```
C       f0::/64[2]
is directly connected, C,FastEthernet0/0
C       f0::26/128[2]
is directly connected, L,FastEthernet0/0
R       17e1::/64[1]
            [120,2] via fe80::2e0:fff:fe9b:7cf1(onFastEthernet0/0)
C       2603::/64[2]
is directly connected, C,Serial0/3
C       2603::26/128[2]
is directly connected, L,Serial0/3
R       26f0:1::/64[1]
            [120,2] via fe80::2e0:fff:fe7a:840(onSerial0/3)
C       fe80::/10[2]
is directly connected, L,Null0
…
```

步骤12：R3 基本配置验证。

```
Router#sho ipv6 rip r1 da
 f0::/64
            [1,120]    via if 2     R    from:[1,0]
 17e1::/64
            [1,120]    via if 3     R    from:[1,0]
 2603::/64
            [2,120]    via fe80::2e0:fff:fe7a:a00 (on FastEthernet0/0)    N, from:
[fe80::2e0:fff:fe7a:a00]
 26f0:1::/64
            [3,120]    via fe80::2e0:fff:fe7a:a00 (on FastEthernet0/0)    N, from:
[fe80::2e0:fff:fe7a:a00]

Router#show ipv6 route
Codes: C - Connected, L - Local, S - Static, R - Ripng, B - BGP
       ON1 - OSPF NSSA external type 1, ON2 - OSPF NSSA external type 2
       OE1 - OSPF external type 1, OE2 - OSPF external type 2
       DHCP - DHCP type

VRF ID: 0
R       2603::/64[1]
            [120,2] via fe80::2e0:fff:fe7a:a00(onFastEthernet0/0)
R       26f0:1::/64[1]
            [120,3] via fe80::2e0:fff:fe7a:a00(onFastEthernet0/0)
C       f0::/64[3]
is directly connected, C,FastEthernet0/0
C       f0::17/128[3]
is directly connected, L,FastEthernet0/0
```

```
C       17e1::/64[4]
        is directly connected, C,Ethernet0/1
C       17e1::17/128[4]
        is directly connected, L,Ethernet0/1
C       fe80::/10[4]
        is directly connected, L,Null0
...
```

任务五 IPv6 OSPFv3 配置

♂ 需求分析

某公司准备对公司网络进行升级，引入新的网络架构——采用 IPv6 来建立公司的网络，工程师通过配置完成了网络的连接。

♂ 方案设计

在两台路由器上启用 IPv6 routing，在接口上配置 IPv6 协议后，通过配置 OSPFv3 相关命令来建立 OSPF 邻居关系，观察学习到的路由。

所需设备如图 10-5-1 所示。
（1）路由器 2 台。
（2）Console 线 1 条。
（3）PC 1 台。
（4）网线若干。

图 10-5-1 IPv6 OSPFv3 配置

任务要求：在任务开始前，路由器为默认配置；按照拓扑图连接网络，在相连的 S 端口上配置 PPP，使串口链路 UP；Router-A 的 Console 口接上 Console 线，并与 PC 相连；在配置 Router-B 的时候，将 Console 线接到 Router-B 的 Console 口上；启动 OSPFv3 进程时会自动寻找路由器上配置的最大 IPv4 地址（loopback 端口优先）作为 router-id，若找不到，则需要手工配置 router-id；全局启动 OSPFv3 进程前应打开 IPv6 unicast-routing；在接口上启动 OSPFv3 实例前请确认端口上启动了 IPv6 协议。

♂ 知识准备

IPv6 中使用地址/前缀长度描述地址，而没有子网掩码的概念。IPv6 中的链路类似于 IPv4 中的子网或网络。OSPFv3 最大的变化就是对 IPv6 地址的支持，以及对 IPv6 体系架构的兼容。

另外，OSPFv3 在 OSPFv2 的基础上，对功能做了增强。OSPFv3 保留了 OPSFv2 的大部分机制。为了支持 IPv6 地址、IPv6 报文结构和体系，OSPFv3 主要做了以下修改。

IPv6 使用链路表示结点赖以在链路层通信的媒介或工具。Interface 连接到链路上。多个 IPv6 地址前缀可以分配到一个单独的链路上；对连接到链路上的两个结点，即使它们的 IPv6 地址前缀不同，也可以直接通信。相应的，OSPF v3 也运行在链路上，而不像 IPv4 一样是基于网段的。链路的概念取代了 OSPF v2 中的网络和子网。因此，OSPF 接口是连接到链路上的而不是子网上的。这一变化影响了 Hello 报文的接收，以及 Hello 报文和 Network-LSA 的内容。

OSPFv3 支持在单链路上运行多实例。这使得多个供应商在共享一台甚至多台网络设备的情况下，仍然可以保持各自网络的独立运行。在 OSPF v2 中，是通过设置不同的 OSPF 验证来实现的。

在单链路上运行多实例是通过在 OSPF 报文头和 OSPF 接口数据结构中包含实例号（Instance ID）而实现的。实例号只影响 OSPF 报文接收。

IPv6 中的本地链路地址用于单链路上的邻居发现、无状态自动配置等。对于以本地链路地址作为源地址的报文，IPv6 路由器不做转发。本地链路单播地址是 FF80/10。

OSPFv3 假定每个路由器的物理接口都分配了本地链路单播地址。除了虚连接以外，所有的 OSPFv3 接口都使用本地链路单播地址作为报文的源地址。路由器从链路上学到其他路由器的本地链路单播地址，再使用这些地址作为转发报文的下一跳。

虚连接使用全球范围地址或本地站点地址作为 OSPF 报文源地址。

本地链路地址只出现在 Link-LSA 中，其他 OSPF LSA 不使用本地链路地址。Inter-area-prefix-LSA、AS-external-LSA 和 intra-area-prefix-LSA 中不能携带本地链路地址。

OSPFv3 本身没有认证功能。因此，OSPFv3 报文头中去掉了 AuType 和 Authentication 字段。相应的，所有的 OSPF 区域和接口数据结构都去掉了认证相关域。

OSPFv3 的认证依赖于 IPv6 报文的认证头和 IP 封装安全有效载荷报头。OSPFv3 通过这些 IP 报文头来确保路由交换的完整性和认证/保密。

OSPFv3 报文利用 IPv6 标准的 16 位完整校验和防止报文数据的随机错误。该校验覆盖了整个 OSPF 报文和伪 IPv6 头。

任务实现

步骤 1：在 Router-A 上启用 IPv6 routing，在端口 S1/1 上配置 IPv6 enable。

```
RouterA_config#ipv6 unicast-routing
RouterA_config_S1/1#ipv6 enable
```

步骤 2：在 Router-A 上验证配置。

```
RouterA#show ipv6 route

Codes: C - Connected, L - Local, S - Static, R - Ripng, B - BGP
       ON1 - OSPF NSSA external type 1, ON2 - OSPF NSSA external type 2
       OE1 - OSPF external type 1, OE2 - OSPF external type 2
       DHCP - DHCP type
VRF ID: 0
C       fe80::/10[1]
    is directly connected, L,Null0
```

```
C       fe80::/64[1]
is directly connected, C, Serial1/1
//接口上自动生成的 linklocal 地址
C       fe80::2e0:fff:fe26:2a58/128[1]
is directly connected, L, Serial1/1
C       ff00::/8[1]
is directly connected, L,Null0
```

步骤 3：在 Router-A 上创建 loopback0 端口，并配置 IPv4 地址；全局启动 OSPFv3 进程。

```
RouterA_config#interface loopback 0
RouterA_config_l0#ip address 1.1.1.1 255.255.255.0
RouterA_config#ipv6 router ospf 1
```

步骤 4：验证配置。

```
RouterA_config_ospf6_1#show ipv6 ospf 1
//选择最大的 loopback 地址作为 OSPF v3 进程的 router-id
Routing Process "OSPFv3 (1)" with ID 1.1.1.1
  SPF schedule delay 5 secs, Hold time between SPFs 10 secs
  Minimum LSA interval 5 secs, Minimum LSA arrival 1 secs
  Number of external LSA 0. Checksum Sum 0x0000
  Number of AS-Scoped Unknown LSA 0
  Number of LSA originated 0
  Number of LSA received 0
  Number of areas in this router is 0
```

步骤 5：在 Router-A 的端口 S1/1 上启动 OSPFv3 进程，并指定区域号。

```
RouterA_config_S1/1#ipv6 ospf 1 area 0
```

步骤 6：验证配置。

```
RouterA_config_ospf6_1#show ipv6 ospf 1
Routing Process "OSPFv3 (1)" with ID 1.1.1.1
  SPF schedule delay 5 secs, Hold time between SPFs 10 secs
  Minimum LSA interval 5 secs, Minimum LSA arrival 1 secs
  Number of external LSA 0. Checksum Sum 0x0000
  Number of AS-Scoped Unknown LSA 0
  Number of LSA originated 2
  Number of LSA received 0
  Number of areas in this router is 1
    Area BACKBONE(0)//启动骨干域
      Number of interfaces in this area is 1
      SPF algorithm executed 0 times
  Number of LSA 1.    Checksum Sum 0xF138
      Number of Unknown LSA 0
RouterA#debug ipv6 ospf packet
//端口 S1/1 上开始发送 Hello 报文
Packet[SEND]: src(fe80:4::2e0:fff:fe26:2a58) ->dst(ff02::5)
OSPFv3 Header
  Version 3   Type 1 (Hello)   Packet length 36
```

```
        Router ID 1.1.1.1
        Area ID 0.0.0.0
        Checksum 0x0000    Instance ID 0
    OSPFv3 Hello
        Interface ID 4
        RtrPriority1    Options 0x000013 (-|R|-|-|E|V6)
        HelloInterval10    RtrDeadInterval 40
        DRouter1.1.1.1BDRouter 0.0.0.0
        # Neighbors 0
```

步骤 7：在 Router-B 上重复上面的 3 个步骤，其中，loopback0 地址配置为 2.2.2.2。

```
RouterB_config#ipv6 unicast-routing
RouterB_config#interface S1/0
RouterB_config_S1/0#ipv6 enable
RouterB_config#interface loopback 0
RouterB_config_l0#ip address 2.2.2.2 255.255.255.0
RouterB_config_l0#exit
RouterB_config#ipv6 router ospf 1
RouterB_config#interface S1/0
RouterB_config_S1/0#ipv6 ospf 1 area 0
```

步骤 8：验证配置。

```
RouterB#show ipv6 ospf neighbor
OSPFv3 Process (1)    //邻居状态为 FULL 状态
Neighbor ID       Pri    State         Dead Time    Interface    Instance ID
1.1.1.1           1      Full/-        00:00:29     Serial/0     0
```

步骤 9：在 Router-A 的 F0/0 端口上配置 IPv6 global 地址 2000::1/64，并在 F0/0 上启动 OSPFv3 进程。此时，应该可以在 Router-B 上学习到 2000::/64 的路由。

```
RouterA_config#interfacef0/1
RouterA_config_f0/1#ipv6 address 2000::1/64
RouterA_config_f0/1#ipv6 router ospf 1
```

步骤 10：验证配置。

```
RouterB_config#show ipv6 ospf database
          OSPFv3 Router with ID (2.2.2.2) (Process 1)
          Link-LSA (Interface FastEthernet0/0)
Link State ID    ADV Router     Age  Seq#        CkSum     Prefix
0.0.0.4          2.2.2.2        1091 0x80000001  0x9d30    0
0.0.0.4          1.1.1.1        545  0x80000002  0xb853    0
                 Router-LSA (Area 0.0.0.0)
Link State ID    ADV Router     Age  Seq#        CkSum     Link
0.0.0.0          2.2.2.2        1086 0x80000002  0xc53b    1
0.0.0.0          1.1.1.1        514  0x80000006  0x9f59    1
              Intra-Area-Prefix-LSA (Area 0.0.0.0)
Link State ID    ADV Router     Age  Seq#        CkSum     Prefix    Reference
0.0.0.1          1.1.1.1        513  0x80000002  0xa2b3    1         Router-LSA
RouterB_config#show ipv6 route
```

```
Codes: C - Connected, L - Local, S - Static, R - Ripng, B - BGP
       ON1 - OSPF NSSA external type 1, ON2 - OSPF NSSA external type 2
       OE1 - OSPF external type 1, OE2 - OSPF external type 2
       DHCP - DHCP type
VRF ID: 0
O       2000::/64[1]//学习到的 OSPF 路由
           [110,20] via fe80:4::2e0:fff:fe26:2a58(onSerial1/0)
C       fe80::/10[1]
is directly connected, L,Null0
C       fe80::/64[1]
is directly connected, C, Serial1/0
C       fe80::2e0:fff:fe26:2d98/128[1]
is directly connected, L, Serial1/0
C       ff00::/8[1]
is directly connected, L,Null0
```

认证考核

实训题

背景与需求：A 公司有 3 台路由器，通过以太网口连接起来，公司决定部署一个 IPv6 的网络，如图 10-5-2 所示，请实现此网络。

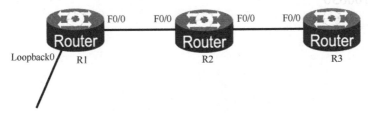

图 10-5-2　A 公司网络拓扑结构

为路由器的接口配置 IPv6 地址，在路由器上启用 IPv6 的 RIP 路由协议，查看如表 10-5-1 所示的路由表。

表 10-5-1　路由表

路由器	路由信息
R1	F0/0　：FEC0:0:0:1001::1/64
	Loopback 0: 1111:1:1:1111::1/64
R2	F0/0　：FEC0:0:0:1001::2/64
	F0/1 : FEC0:0:0:1002::1/64
R3	F0/1 : FEC0:0:0:1002::2/64

完成标准：网络接口地址配置正确，网络的连通性可以得到证实，路由协议配置正确，查看路由表。

反侵权盗版声明

电子工业出版社依法对本作品享有专有出版权。任何未经权利人书面许可，复制、销售或通过信息网络传播本作品的行为；歪曲、篡改、剽窃本作品的行为，均违反《中华人民共和国著作权法》，其行为人应承担相应的民事责任和行政责任，构成犯罪的，将被依法追究刑事责任。

为了维护市场秩序，保护权利人的合法权益，我社将依法查处和打击侵权盗版的单位和个人。欢迎社会各界人士积极举报侵权盗版行为，本社将奖励举报有功人员，并保证举报人的信息不被泄露。

举报电话：（010）88254396；（010）88258888
传　　真：（010）88254397
E-mail： dbqq@phei.com.cn
通信地址：北京市万寿路 173 信箱
　　　　　电子工业出版社总编办公室
邮　　编：100036